Oxford Applied Mathematics and Computing Science Series

General Editors
J. N. Buxton, R. F. Churchhouse, and A. B. Tayler

OXFORD APPLIED MATHEMATICS AND COMPUTING SCIENCE SERIES

Jagdish J. Modi

University of Cambridge

Parallel Algorithms and Matrix Computation

CLARENDON PRESS · OXFORD

1988

Oxford University Press, Walton Street, Oxford OX2 6DP

Oxford New York Toronto
Delhi Bombay Calcutta Madras Karachi
Petaling Jaya Singapore Hong Kong Tokyo
Nairobi Dar es Salaam Cape Town
Melbourne Auckland

and associated companies in
Berlin Ibadan

Oxford is a trade mark of Oxford University Press

Published in the United States
by Oxford University Press, New York

© *Jagdish J. Modi, 1988*

British Library Cataloguing in Publication Data
Modi, Jagdish
Parallel algorithms and matrix
computation.
1. Computer systems. Parallel-processor
systems. Programming. Algorithms
I. Title II. Series
005.1
ISBN 0-19-859655-3
ISBN 0-19-859670-7 (pbk.)

Library of Congress Cataloging in Publication Data
Modi, Jagdish.
Parallel algorithms and matrix computation/Jagdish Modi.
(Oxford applied mathematics and computing science series)
Bibliography. Includes index.
1. Parallel processing (Electronic computers) 2. Algorithms.
3. Matrices. I. Title. II. Series.
QA76.5.M543 1988 004'.35—dc 19 88-9687
ISBN 0-19-859655-3
ISBN 0-19-859670-7 (pbk.)

Typeset in Northern Ireland by The Universities Press (Belfast) Ltd
Printed in Great Britain
at the University Printing House, Oxford
by David Stanford
Printer to the University

This book is dedicated with
affection and respect to my parents,
my sister, and my younger brother

Preface

The introduction of parallel computers some twenty years ago both provided an opportunity and prompted a need to develop parallel algorithms. In the process of this development, new perspectives have opened up, and a general theory of parallel algorithms, connecting certain established principles in logic and mathematics, will undoubtedly emerge. The elaboration of parallel methods has permitted not only more freedom of expression in problem analysis, but more generally has stimulated wide-ranging intellectual inquiry. Advanced applications, including image processing, robotics, speech recognition, and artificial intelligence, which have overtaken the capacity of the older sequential machines, can expect to benefit from this scientific impetus.

The purpose of this book is to bring together and further articulate the fundamental concepts in parallel computing. I will be focusing particularly on numerical linear algebra in view of its primary importance to both industry and academic research. I have divided the book into two parts. The first part (Chapters 1 to 3) discusses general principles and algorithmic techniques, treating a number of examples, and contains the essentials of a final-year undergraduate course in computer science or computational mathematics. This is complemented in Part 2 (Chapters 4 to 7) by a detailed exposition of some of the key areas of application in linear algebra. While the first part of the book will be of interest to a wide audience including those acquainted only with serial computing, the second part will demand a level of mathematical maturity typical of a graduate in mathematics, physical sciences, or engineering.

The book is structured in the following way. Chapter 1 introduces the fundamentals of parallel computing. Here I provide an up-to-date account of, *inter alia*, the classification of parallel computers, the interplay between machine architecture and algorithm design, performance measurement, and the classification of parallel algorithms. Chapter 2 puts the emphasis on algorithmic design and the variety of newly discovered techniques which fully exploit the features of parallel computers. Chapter 3 is devoted entirely to parallel sorting techniques, demonstrating how parallelism may be incorporated into some already well-established serial methods.

The remaining chapters concentrate on matrix computations. The discussion of the solution of linear algebraic equations (Chapter 4) is limited to a selection of methods designed to illustrate research into the use of parallelism. In Chapter 5, the Jacobi method for the symmetric eigenvalue problem is discussed in detail. The remarkable stability of the orthogonal transformations, together with the simultaneous application of independent plane rotations, leads to a consideration of these techniques in other applications, notably QR factorization (Chapter 6) and singular-value decomposition and the linear least-squares problem (Chapter 7).

Throughout the book, I have tended to stress the identification of parallelism rather than its implementation on particular architectures, though numerous results, especially in Part 2, apply to the Distributed Array Processor (DAP).

It is a pleasure to acknowledge the assistance of many friends and colleagues. I express my sincere thanks to Professors D. J. Wheeler and D. Parkinson and Dr R. K. Livesley for providing constructive criticism and invaluable guidance. I am grateful for their detailed comments on specific chapters to Professors J. K. Iliffe, L. M. Delves, R. O. Davies, C. C. Paige, Drs J. S. Rollett, P. M. Flanders, M. A. Sabin, and Mr S. J. Hammarling.

I am indebted to Drs H. R. Pratt and E. J. Primrose for greatly enhancing my text, and to Mrs M. Ward for—with immense patience—typing the manuscript. I would also like to express my appreciation to the Oxford University Press, and in particular to Mr J. Bentin, for their consistent co-operation. Last but not least I would like to thank my parents, whose unfailing generosity I can never hope to repay.

Cambridge　　　　　　　　　　　　　　　　　　　　　　　　J.J.M.
February 1988

Contents

Part 1

Fundamentals of Parallel Computation

1 General principles of parallel computing

1.1 Parallel computing, past and present

The 1980s will go down in history as the decade in which parallel computing first had a significant impact in the scientific and commercial worlds. A substantial improvement in speed over serial computation has already been secured, by up to a factor of 1000, for a number of major applications, but the scope for further developments remains considerable. Machines are currently being built that exhibit an extensive degree of artificial intelligence, and incorporate a significant amount of parallelism. For example, the Connection Machine, based on learning algorithms, has 65,536 intelligent memory cells capable of communicating with other cells and thus representing a semantic network. As technological sophistication increases, more complex systems aimed at special-purpose applications such as the processing of speech and natural language will undoubtedly emerge. The concepts of both parallel computation and artificial intelligence are, however, far from original, and date back to the earliest developments in computing.

In an often-quoted statement, Lady Lovelace (1843), discussing Babbage's machine, wrote that 'The Analytical Engine has no pretensions to *originate* anything. It can do *whatever we know how to order it to perform*' (her italics). But she was also well aware that it can execute conditional instructions and therefore adapt its own operation during the course of a calculation: 'The engine is capable, under certain circumstances, of feeling about to discover which of two or more possible contingencies has occurred, and of then shaping its future course accordingly.' Menabrea (1842), in his *Sketch of the Analytical Engine Invented by Charles Babbage,* stresses accuracy, rapidity, and the saving of intellectual drudgery as the advantages of the machine. But Turing (1950) pointed out that in fact 'the Analytical Engine was a universal digital computer', and therefore with adequate storage capacity and speed could mimic any machine, even a hypothetical one that 'thinks for itself' (although, as he said: 'Probably this argument did not occur to the Countess or to Babbage'). In this celebrated paper *Computing Machinery and Intelligence,* Turing essentially gives an affirmative answer to the question: 'Can machines think?'

In Menabrea's report we also read that 'when a long series of identical computations is to be performed, such as those required for the formation of numerical tables, the machine can be brought into play so as to give several results at the same time, which will greatly abridge the whole amount of the processes'. It appears that Babbage was conscious of the basic speed advantage parallelism can offer. But if the concept of parallelism is far from novel, no more than a very limited amount—in the form of instruction pipelining—was envisaged in the classical Von Neumann concept of the serial computer and incorporated in the earliest machines such as EDSAC 1 (1949) and UNIVAC (1951). The application of parallelism at various levels (job level, algorithm level, statement level, etc.) is comparatively recent.

Figure 1.1 shows the increase in arithmetic speed through the development of serial computers since 1950. Improvements in hardware have brought tremendous advances, for example the successive replacement of vacuum-tube switches by transistors, large-scale integrated circuits (LSI), very large-scale integrated circuits (VLSI), and ultra large-scale integrated circuits (ULSI). By the late 1970s, although the transmission of electronic signals through wires had become so fast—almost approaching the speed of light—this nonetheless proved to be the limiting factor compared with electronic switch speed. Minimizing the length of wire unfortunately heats up the switches, and consequently taxes the physical limitations of thermal conductivity. The Cray-1 *vector processor,* which is discussed later, opted for small-scale integration with an elaborate cooling system to prevent meltdown. A twenty-fold speed-up every decade (Fig. 1.1) has thus resulted primarily from progress in hardware technology and—to a lesser extent—from slave processing, interleaving of memory, and the incorporation of parallelism at lower levels.

With the increase in computing capability, certain large-scale computational problems, especially those concerned with pre-

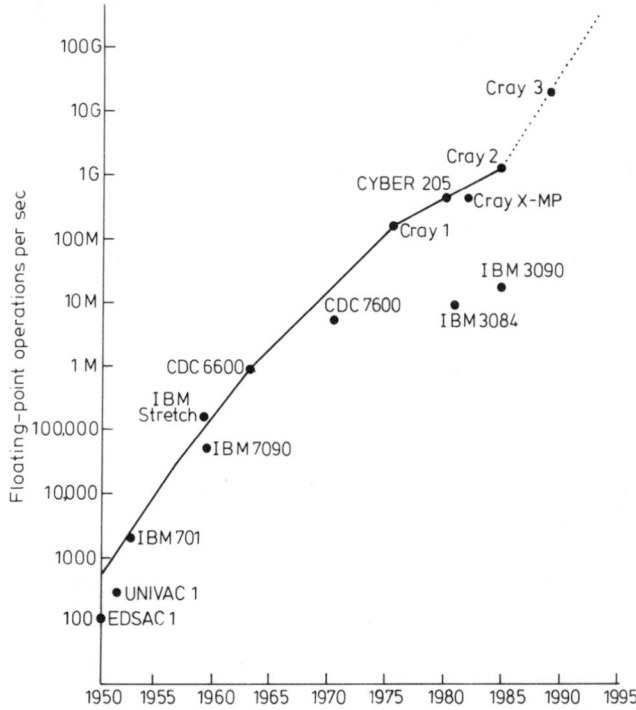

Fig. 1.1   Improvement in peak computer performance 1950–1995

dicting the behaviour of complicated physical systems by means of mathematical modelling, began to approach feasibility, and demands for even greater capacity became more pressing. In fact, this state of affairs is self-perpetuating: more computing power is always going to lead to the demand for yet more computing power. At any given moment, the capabilities of contemporary computers are inevitably one step behind the perceived needs for scientific and technological applications.

While machines of the classical Von Neumann design, with minor modifications, have thus been made very much faster, it seems certain that, to obtain comparable improvements in speed in the future, it will be necessary to incorporate a high degree of parallelism. Speed improvement on serial machines is constrained by an inability to increase the transfer speeds of operands to and from the data buffers and the functional units. Particularly when lengthy repetitive calculations are required, the cost associated with data transfer becomes considerable: main storage devices currently in use, for instance, have access times of 100–150 nsec, and longer if accessed through a shared highway, whereas typical modern processor times are of the order of 50 nsec. The objective is therefore to strike an optimal balance between processor speed and memory access rate.

One well-established method of improving processing speed, by means of limited parallelism at assembler level, is to start each operation before the previous one is completed. This approach, known as pipelining, is discussed in detail in Section 1.9. Machines like those of the Cray and Cyber series couple the technique of pipelining with independent hardware units for performing distinct operations such as addition and multiplication. These allow a greater number of simultaneous operations, which in turn result in greater speed. The term vector processor is commonly used to describe such a system.

In the 1980s, VLSI technologies have led to the development of multiprocessors. High-speed buffers may be omitted entirely, and a large number of processing units connected directly to the memory banks. Memory is distributed among these processors, with the possible inclusion of a global memory to which all units have access; processor-interconnection schemes based on hypercubes, rings, and lattices have also been designed (Section 1.7). This concept is the basis of parallel processing. The single central processing unit of the classical Von Neumann design is replaced by many processors, which, even if individually somewhat slower, accelerate the speed of processing by operating in parallel. In terms of cost, an n-processing-unit parallel system costs much less than n times a single-unit system. These developments in hardware design were motivated by the use of semiconductor stores as memory components. Semiconductor devices have the same physical and electrical properties as logical devices, and it may be appropriate to consider the memory units and processing units together as a single entity: the 'active memory array' in the phrase coined by Iliffe (1982). A number of parallel machines have been built that incorporate vector processing, multiprocessing, or a combination of both. Figure 1.2 shows the three generations of supercomputers since the 1970s,

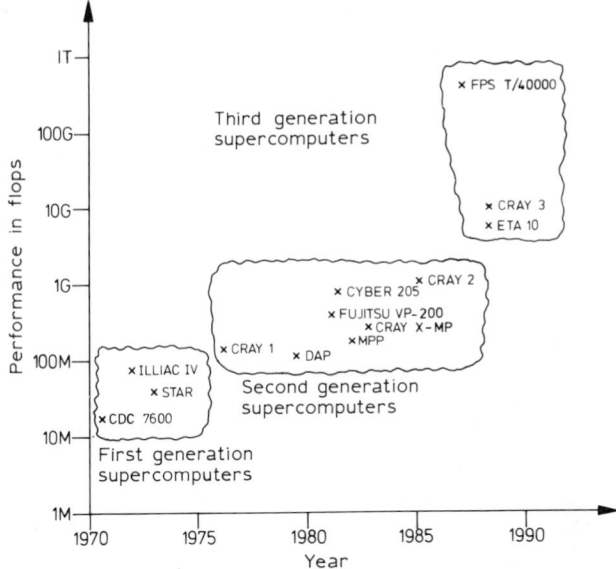

Fig. 1.2 Generations of supercomputers since the 1970s. The figures show approximate performance only, and strongly depend on the particular application

in terms of peak performance measured in *flops* (floating-point operations per second).

The innovation of parallel computing has also added a new dimension to the design of algorithms and programs. Parallel programming is not a simple extension of serial programming. In order to exploit the possibilities offered by parallelism, programmers need to 'think in parallel' and reconsider the solution process. Experience has shown that judgements of efficiency based on serial techniques can easily be overturned in a parallel environment (Section 1.10). Algorithms that are obsolete so far as implementation on serial machines is concerned often show a high degree of parallelism—i.e. contain numerous subcalculations that are independent of one another and may therefore be executed simultaneously—and can outperform the best serial techniques (for example, the Jacobi method for the eigenvalue problem; see Chapter 5). But, in addition to demanding new techniques of algorithm design, the introduction of extended parallelism also requires us to rethink what we know about systems, parallel languages, non-numerical problems, and numerical methods in general.

This chapter is divided as follows. Initial sections provide background concepts with an informal survey of possible architectures and discuss Flynn's classification of parallel computers. The introduction of parallelism into computer systems has led to various architectures consisting of processing elements, and the interconnections between them. To discuss the issues related to transmission between individual processing elements we need some formal algebra, which is provided by graph theory (Section 1.6). We describe some commonly available networks (Section 1.7), particularly with respect to symmetry, homogeneity and embedding (Section 1.8). Although a vector processor does not constitute a truly parallel system, it nevertheless provides a significant improvement in speed (Section 1.9). The characteristics of parallel computers and parallel algorithms are qualitatively different from those of serial computation, and these issues are dealt with in later sections. The second part of the chapter thus provides a formal study, and leads naturally into a discussion of techniques for developing parallel algorithms in Chapter 2.

1.2 Classification of parallel computers

In a situation of rapid development in parallel computing, it is hardly surprising that no entirely satisfactory taxonomy of machines has yet been established. With some justification, it has been suggested that Flynn's classification, dating as it does from 1966 when parallel computing was in its infancy, does not adequately reflect current architectural designs. Kuck (1982) has attempted to provide a more up-to-date machine taxonomy and the reader is referred to his paper for further discussion of this subject. Nevertheless, as a guideline, Flynn's categories remain useful. He classifies computers into four types:

1. SISD: single instruction stream, single data stream
2. SIMD: single instruction stream, multiple data stream
3. MISD: multiple instruction stream, single data stream
4. MIMD: multiple instruction stream, multiple data stream.

These are displayed in Fig. 2.1. The term SISD essentially designates the classical serial machine design. Each of the others refers to a machine with a number of processors operating in parallel, to which multiple instruction and/or data streams are directed. For a discussion of MISD design the reader is referred to Flynn's paper. 'Multiple data stream' is now considered the most appropriate description for existing parallel machines. Table 2.1 groups some of these into SIMD, MIMD, and vector processors; the latter can perhaps best be thought of as SISD machines incorporating some MI and MD features at a low level.

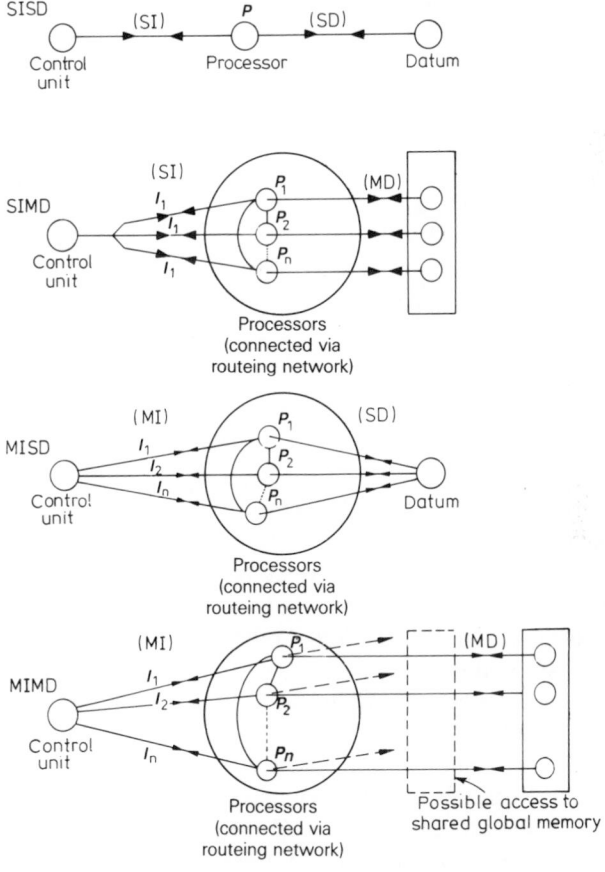

Fig. 2.1 Flynn's classification of computers

Table 2.1   A spectrum of supercomputers

Supercomputers	First announced	Class of machine	Number of processors and topology
Star 100	1971	Vector capability	4
Cray 1	1976	Vector capability	1
Cyber 205	1980	Vector capability	1
Cray X-MP 2	1982	Vector capability	1,2,4
Fujitsu VP-200	1982	Vector capability	1
Cray 2	1985	Vector capability	4
Eta 10	1987	Vector capability	2,4,6,8
Cray 3	1988	Vector capability	16
ILLIAC IV	1972	SIMD	64, linear
DAP (Bit oriented)	1974	SIMD	4096, lattice
Connection machine (Bit oriented)	1985	SIMD	65 536, hypercube
BBN butterfly	1981	MIMD	256, butterfly
Cyberplus	1983	MIMD	16, ring
Intel hypercube	1985	MIMD	128, hypercube
Ncube	1985	MIMD	1024, hypercube
Meiko surface	1986	MIMD	40, reconfigurable
FPS T/40000	1988	MIMD	16 384, hypercube

In addition to these there are a number of research MIMD machines:

the *Cedar* machine, at the University of Illinois, has 32 processors with a multistage crossbar;

the *Ultracomputer* (proposed), at New York University, has 4096 processors with an omega interconnection;

the *Alice* machine, at Imperial College, has 64 processors with a shuffle-exchange interconnection;

the *Cosmic cube,* at Caltech, has 64 processors with a hypercube interconnection;

the *Grip* machine, at University College, London, has 120 processors with a bus interconnection;

the *Field analysis* machine, at Cambridge University, has 64 T800 transputers (having integral floating-point hardware) with a hamiltonian interconnection and a crossbar switch.

An SIMD system comprises a master control unit and a number of identical processing elements. The control unit transmits the same instruction stream to each of the processing elements. An MIMD system is more general, in that individual processors may be instructed to perform different operations: one may do addition while another does multiplication. Both SIMD and MIMD systems are derived from the simple idea of trying to increase speed by multiplying the number of processing elements. A less radical—but also more limited—approach to high speed computing is through the pipelining and chaining available on vector processors, mentioned in Section 1.1 and described in more detail in Section 1.9. These techniques are attractive to users primarily because compilers can be made to recognize certain types of parallelism on existing software; relatively little program modification is required, and much of it at the level of coding rather than algorithm design.

Farms and cubes

Another simple and popular classification of parallel processors is to divide them into two broad categories: farms and cubes. A typical farm comprises a small number of processors, each comparable to a minicomputer in performance, as typified, for example, by the Alliant FX/series with eight processors. The processors communicate through a shared memory which is accessed via a bus or through the routeing network connecting certain processors directly to others. Each processor is autonomous and may operate as a stand-alone computer if needed. Because of the relatively small number of processors, usually about ten, the specification of parallelism onto such systems requires minimum modification to sequential programs. Furthermore, because each processor is fairly complex, capable of performing numerous calculations, the system is best suited to task parallelism, where the problem is broken down into a small number of tasks, and these are then farmed out onto cooperating processors and processed simultaneously. Special compilers already exist which can, to a limited extent, recognize task parallelism in existing sequential programs, particularly those written in Fortran. With respect to granularity, a function of the number of processors and their complexity (see Chapter 2, Section 2.9), a farm closely resembles a coarse-grain system.

A cube, on the other hand, characterizes a machine having a large number of processors, ranging from single-bit processors, as in the connection machine, to moderately complex ones, as in the Intel hypercube, though generally less complicated than a farm processor. The processors are usually connected by a routeing network. Currently available systems have between 128 and 65 536 processors with varying degrees of complexity. A number of such systems have routeing networks forming a hypercube, or some other topological connectivity with similarly desirable features (see Section 1.7). The processors are usually built from microprocessor chips (such as the transputer) with additional memory and arithmetic units. However, unlike in a farm, processors are not autonomous and cannot be utilized as stand-alone micros, but need to be driven by a separate host computer. Because of the large number of processors at the programmer's disposal, the use of the cube requires a careful analysis of the problem. It is often necessary to start from scratch, and to develop new algorithms that make efficient simultaneous use of the processors, at minimum communication costs. Such systems have led to the exploitation of algorithmic parallelism. In this book, we will almost exclusively discuss special techniques and algorithms suited to this class of machine. With respect to granularity, the cube closely resembles a system having fine-to-medium grain.

Integrated memory v. shared memory

This classification, suggested by Gurd (1988), is a refinement of Flynn's. Rather than concentrate on the number of active instruction streams, the emphasis is put on the relationship between processing elements, memory module, and interconnection network. Two configurations are commonly found in practice: those with *integrated memory,* and those with *shared memory.* In the former case, there is no memory in the system other than that which is locally addressable within the processing elements. In the latter case, there is a central memory (global or shared) that can be addressed from any of the processing elements, irrespective of whether or not they contain their own locally addressable memory. Issues such as which computational model is being implemented, or whether there is a separate instruction stream, are given secondary importance. Thus a revised taxonomy of parallel machines divides machines into those with integrated memory and those with shared memory, and then applies Flynn's classification to each of these subgroups. This links machines with similar performance characteristics more closely together. Further details may be found in Gurd (1988), where the author applies this taxonomy to a number of currently available systems.

In Sections 1.3 to 1.4 we shall continue our discussion of Flynn's classification, and then proceed to examine interconnection networks and their properties in Sections 1.5 to 1.8.

1.3 Single instruction multiple data (SIMD) architecture

The master control unit (MCU) of an SIMD machine broadcasts a stream of identical instructions to each of the processing elements (see Fig. 3.1). These instructions are then executed at the same time: that is, the operations are synchronous.

Each processing element has a private memory, and some systems may also include access to a global memory. When the processing element requires data that are not stored in its own memory, then some means of connecting the processing elements is needed in order to effect the necessary transfer. The repositioning of data items can often be done in parallel through the use of a routeing network which links certain processing elements directly to others. Various topologies for this are described in Section 1.7.

Each of the processing elements may operate on a word (typically 32 or 64 bits) or on single-bit operands during each memory cycle, and this feature distinguishes two types of SIMD

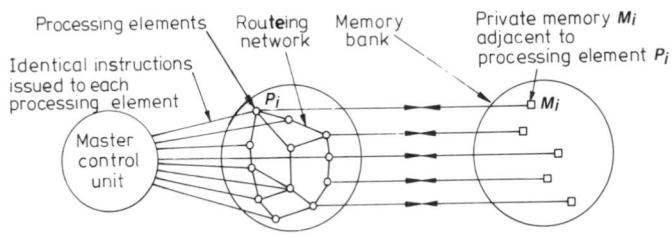

Fig. 3.1 SIMD design

system: word-organized and bit-organized. The arithmetic operations within a word-organized system are similar in principle to those found in serial machines, and will therefore not be described here. Bit-organized machines, on the other hand, have greatly influenced the design of parallel algorithms: the algorithm for extrema, for example, depends on the bit length of the numbers and not on the size of the set, provided that enough processing elements are available (see Chapter 2, Section 2.7).

Bit-organized array processors

The use of the term 'array processor' is frequently ambiguous: it may refer to an SIMD system, an MIMD system, or a vector processor. Here it describes an SIMD system in which the routeing network plays an important role. Array processors generally incorporate a large number of identical processing elements connected in a particular topology. These act synchronously under the control of a single unit issuing a set of identical instructions. For the case of the distributed array processor (DAP), the arithmetic within each processing element is carried out as follows. In Fig. 3.2, A, Q, and C are all single-bit registers, and the most complex instruction uses the one-bit full adder to sum the contents of Q register and C register and a selected bit from store to leave the two-bit result in Q and C. It is

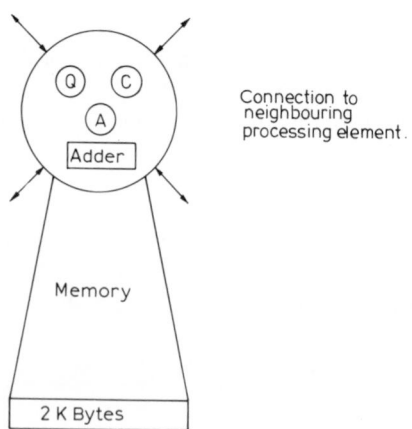

Fig. 3.2   Processing element of DAP

almost invariably necessary to limit the operation to only certain of the processing elements. The activity register *A*, through which the programmer may selectively inhibit any element, thus becomes one of the most important features of this architecture.

Each processing element operates on single bits of data from its private memory. Consequently, several instructions may be needed to create even the simplest of arithmetic operations. The computation time for a single operation is therefore a function of the bit complexity of that operation, and this feature has led to a number of interesting algorithms (see Chapter 2, Sections 2.7 and 2.8).

Compared with word-organized systems, bit-organized array processors have a long arithmetic time (for example, 20 µsec for 16-bit integer addition on the DAP), and the overall improvement in speed is achieved not necessarily through making individual processing elements work faster, but through using many of them simultaneously. The speed of any given algorithm on an array processor is also influenced by the routeing network, which is discussed in Section 1.5.

In conclusion, and in summary, the following are some of the advantages of the bit-organized array processor. Firstly, because the processing elements act on single bits of data, variable-length arithmetic is possible, and consequently lower-precision arithmetic is faster than higher-precision arithmetic; further, the data need not be packed and unpacked into artificial word sizes, as in word-based machines, which saves both time and storage space. In the implementation of numerical functions, such as square root, one can take advantage of the bit operation of the machine to produce algorithms that are only marginally more expensive than multiplication (see Chapter 2). Secondly, bit-organized array processors facilitate logical operations, which provide a cheap method of selectively inhibiting the processing elements. In word-organized systems, selection is as costly as arithmetic, and algorithms must therefore use a minimal number of selection operations.

1.4 Multiple instruction multiple data (MIMD) architecture

The MIMD architecture is designed so that each individual processor can execute a separate instruction on a separate item of data (see Section 1.2). It might be thought of as a system of interconnected conventional processors, where each processor can operate independently on its own data stream: in a three-processor system, for example, one of them could perform addition, another subtraction, and the third comparison. Further, since the instructions are independent of each other, the execution of one instruction does not influence the execution of any other, with the consequence that all processors may be operational at all times. Figure 4.1 shows a more detailed configuration of an MIMD system than displayed in Fig. 2.1.

Unlike the SIMD system, the MIMD operates in an asynchronous mode: each processor has its own local memory, arithmetic unit, and instruction counter, but they may communicate with each other, as in the SIMD, through a routeing network, and again the proper choice of configuration can have a significant effect on the speed of execution. In order to discuss the transmission of data between individual processing elements, we need some formal algebra, which is provided by graph theory. We discuss this in Section 1.6, and then consider some commonly available networks (Section 1.7), particularly with respect to *symmetry, homogeneity,* and *embedding* (Section 1.8). These issues are relevant to both SIMD and MIMD systems.

In most MIMD systems, each processor has access to a global memory, which may reduce processor communication delay. In addition, each possesses a private memory which assists in

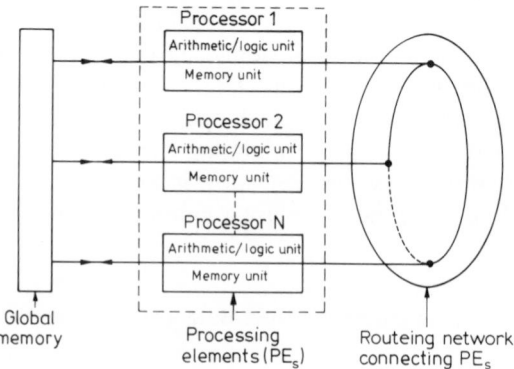

Fig. 4.1 MIMD design

avoiding memory contention (the necessity for queuing when several processors need to draw data from the global memory). In current MIMD systems the number of processors is smaller than in SIMD systems: the MIMD machine Ncube with hypercube configuration, for example, has dimension 10, i.e. $2^{10} = 1024$ processors, and an array of transputers the same number, whereas the SIMD machine DAP with lattice configuration has $64 \times 64 = 4096$ processors, the Connection machine with hypercube topology has $2^{16} = 65\,536$ processors, and the next version with $2^{20} = 1\,048\,576$ processors is already being planned!

It is very likely that, in the near future, small MIMD systems, with a limited number of processors, will be built with complete connectivity, that is, each processor connected to every other one. Systems with a large number of processors are likely to have some direct and some indirect connections via a 'switchboard' (Section 1.5). The impact of these developments on computing speeds would be considerable and would constitute a major breakthrough.

1.5 Networks for processor connectivity

In order to exchange data between different processing elements, some means of communication is necessary. This may be effected by a routeing network (as indicated above, elements may also share a global memory). Alignment of data with elements is carried out in parallel, to the extent that the algorithm and the nature of the routeing network permit. In practice, many particular features of program and algorithm design arise out of limitations imposed by the network.

For example, on a parallel computer having $2n$ elements with only the linear connections $P_1 - P_2 - \cdots - P_{2n-1} - P_{2n}$, a natural method of sorting $2n$ numbers into increasing order is the *odd–even transposition sort*. Assuming the numbers in $P_1, \ldots, P_{2n}$ are stored as $x_1, \ldots, x_{2n}$, the method proceeds as follows.

Step 1 Compare x_{2i-1} and x_{2i}, and store the smaller and larger in P_{2i-1} and P_{2i} respectively for all $i = 1, \ldots, n$ simultaneously; that is,

$$x_{2i-1} \leftarrow \min\{x_{2i-1}, x_{2i}\} \quad \text{and} \quad x_{2i} \leftarrow \max\{x_{2i-1}, x_{2i}\}.$$

Step 2   Compare x_{2i} *and* x_{2i+1} and store the smaller and larger in P_{2i} and P_{2i+1} respectively for all $i = 1,..., n-1$ simultaneously.

Repeat Steps 1 and 2 alternately.

In $2n$ steps, the sorting is accomplished, and it can be shown that no other implementation can succeed in sorting all initial sequences in fewer steps, using linear interconnection of processors. On the other hand, for more extensive routeing networks, there exist sorting algorithms requiring only approximately $\frac{1}{2}(1 + \log_2 n)^2$ parallel steps.

In an ideal situation, each element would be connected to every other one. In most existing systems, however, there are only a reduced number of interconnections, and this ideal is not possible unless a general switching network is used. When the machine is being designed, judgement about a particular connectivity is often based on the class of problem for which the system is intended, and on limitations on the number of connections available from each individual processing element. Reconfigurable connectivity is now available, so that a variety of configurations may be constructed and altered in the course of a computation. Nevertheless, this is likely to be costly in terms of time; thus, in the course of any one calculation, the configuration will remain stable for long periods.

If the connectivity requirement of a particular algorithm does not match the routeing network, then the resulting communication delay can adversely affect processing speed. Kung and Stevenson (1977) have carried out research into software techniques for reducing the routeing time on a parallel computer with a fixed interconnection network.

Routeing networks may be classified as follows.

(a)   Each element connected to all the others directly: this design is not practical at present, other than for exceptionally small machines, because of high wiring costs and complexity.

(b)   Each element connected to all the others with intermediate connections, for example through a 'switchboard': this guarantees that few steps are required to connect any two processors; but some direct connections may nevertheless be advantageous, particularly between near-neighbour processors which are utilized by a number of parallel algorithms (Fig. 5.1).

(c)   Each element connected to a selected number of others.

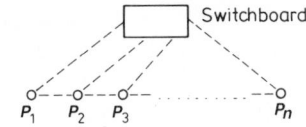

Fig. 5.1 Interconnection via switchboard

A number of different processors of type (c) are currently in use, and others are being planned. We shall discuss some of the more important networks and give a simple example of a problem for which each is ideally suited. But in order to discuss various topologies we need the formal algebra provided by graph theory, which is discussed below.

1.6 Graphs

The natural language for describing networks is that of graph theory, and we therefore give a few relevant definitions. More details can be found in standard texts such as those of Harary (1969) or Bollobás (1979); Wilson (1972) is a short and very readable introduction. Unfortunately, no two authors in this field seem to use exactly the same terminology.

A (simple) *graph G* consists of a finite set of *vertices* (nodes) together with *edges* joining certain pairs of distinct vertices. If we specify an edge by this pair, then we can regard G as $\langle V(G), E(G) \rangle$ where $V(G)$ is a finite set and $E(G)$ is a collection of 2-element subsets of $V(G)$. We denote the edge joining the vertices v_1 and v_2 by $v_1 v_2$. The *size* $|G|$ is the number of vertices. Figure 6.1 shows four graphs of size 4 and two of size 5.

No vertex is joined to itself, and two vertices are joined by at most one edge. If $|G| = n$, and all possible joins occur, we have the *complete graph* K_n on n vertices: thus G_4 above is K_4. The *degree* (valency) of a vertex is the number of edges issuing from it. Thus G_5 has one vertex of degree 4 and four of degree 2. In K_n, each vertex has degree $n - 1$.

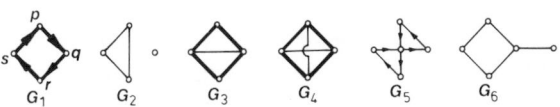

Fig. 6.1 Graphs

In a graph, a *path* of length $n - 1$ is a sequence of n vertices $v_1, ..., v_n$ for which $v_1v_2, ..., v_{n-1}v_n$ are distinct edges; it is a *chain* if the vertices are distinct except, possibly, for $v_1 = v_n$, which makes the path (or chain) *closed*. A closed chain with at least three edges is a *circuit*. A graph is *connected* if any two points can be joined by a path. All the graphs in Fig. 6.1 are connected except for G_2. The *diameter* of a connected graph is the maximum possible 'distance' between any two vertices (length of shortest path joining them): thus G_4 has diameter 1, the graphs G_1, G_3, and G_5 have diameter 2, and G_6 has diameter 3, while the diameter of G_2 is undefined because it is not a connected graph. In practice, it is clearly desirable to have an interconnection of processors with minimum diameter in order to reduce the cost of communication between processors.

If a graph contains a closed path that includes every edge or every vertex exactly once, then it is called *Eulerian* or *Hamiltonian* respectively: for example, G_1 is both Eulerian and Hamiltonian—this is precisely the case for every *cycle*, consisting of a single circuit with no additional vertices or edges. The graphs G_3 and G_4 are Hamiltonian but not Eulerian, while G_5 is Eulerian but not Hamiltonian, and G_6 has neither property. In Fig. 6.1, Hamiltonian paths are indicated by thickened edges and Eulerian paths by arrows (the arrows are not to be confused with the notation for directed graphs, which are discussed later). A connected graph G_n is Eulerian if and only if the degree of every vertex is even. If G_n is a graph of size n, and the degree of every vertex exceeds $\frac{1}{2}n$, then G_n is Hamiltonian (a result due to Dirac).

Two graphs are *isomorphic* if we can set up a 1–1 correspondence (isomorphism) between their vertices with the property that two vertices are joined by an edge in one graph if and only if the corresponding vertices are joined by an edge in the other. Thus, if $V(G) = \{v_1, ..., v_n\}$ and $V(H) = \{w_1, ..., w_n\}$, then the correspondence taking v_1 to w_1, v_2 to w_2, and so on is an isomorphism if v_iv_j is in $E(G)$ just when w_iw_j is in $E(H)$. To put it another way, both graphs can be represented by the same diagram of the type of Fig. 6.1.

A graph may admit *automorphisms* (isomorphisms with itself) besides the identity automorphism (under which each element corresponds to itself). For example, in G_1 the vertices *pqrs* can

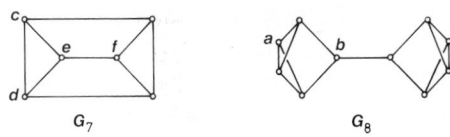

Fig. 6.2 Homogeneity and nonhomogeneity

correspond respectively to *psrq, qrsp, qpsr, rspq, rqps, spqr,* or *srqp* as well as to *pqrs*: so that there are eight automorphisms altogether. For the complete graph K_n, all 1–1 correspondences between $V(K_n)$ and $V(K_n)$ are automorphisms (there are $n!$ of them). A graph with many automorphisms may be described as having a high degree of *symmetry*.

If a graph G has the property that for each pair of vertices v and w there exists an automorphism taking v to w, then we may term G *homogeneous*. Roughly speaking, the graph 'looks the same' from each point. For example, G_1 and G_4 are homogeneous, and also the graph G_7 in Fig. 6.2. A homogeneous graph must be *regular*; all vertices have the same degree. However, this condition is not sufficient, even in a connected graph: for example, the graph G_8 in Fig. 6.2 is regular and connected but not homogeneous (no automorphisms can take a to b because a suitable cut on an edge issuing from b disconnects the graph but this is not true for a).

If the graph 'looks the same' from any one *group* of points related in a certain way and any other group of the same number of points related in the same way, we may regard this as an additional degree of homogeneity. For example, the graph G_1 has the property that, for any two adjacent points, there is an automorphism taking them to any other two adjacent points. We may regard G_1 as more homogeneous (at any rate, in this one respect) than the homogeneous graph G_7 of Fig. 6.2, in which the pair (c, d) cannot be made to correspond to the pair (e, f) because the edge cd forms part of a triangle whereas the edge ef does not.

The routeing network for a parallel computer connects the processing elements by wires (or their equivalent), and in practice each wire can pass data in only one direction: the wire is thus *directed* from one processing element P_1 to another P_2. In most machines there will also be a wire directed from P_2 to P_1, and then it is convenient to regard the pair of wires as forming a single connection which can pass data in both directions. However, certain algorithms do not need both data movements, and machines in which certain connections are unidirectional have been proposed: the appropriate language is now that of directed graphs.

By a *directed graph* or *digraph D* we mean a finite set of vertices, together with *arcs* (directed edges) joining certain pairs of vertices (not necessarily distinct), along which a direction is given. Thus D is $\langle V(D), A(D) \rangle$, where $V(D)$ is a finite set and $A(D)$ is a collection of ordered pairs of elements of $V(D)$. Figure 6.3 shows a number of directed graphs.

The arcs in D_1 are pp (a *loop*), pq, qr, and rq. If v is a vertex of D, then the *out-degree* of v, denoted by $\bar{\rho}(v)$, is the number of arcs of D of the form vw; similarly the *in-degree* of v, denoted by $\bar{\rho}(v)$, is the number of arcs of D of the form wv. In Fig. 6.3, for D_2, we have $\bar{\rho}(q) = 2$ and $\bar{\rho}(q) = 1$.

In a directed graph D, a (directed) *path* of length $n - 1$ is a sequence of n vertices $v_1, \ldots, v_n$ for which $v_1v_2, \ldots, v_{n-1}v_n$ are distinct arcs (we may write $v_1 \rightarrow v_2 \cdots \rightarrow v_n$), and terms such as chain and circuit extend in an obvious way; D is called *strongly connected* if, for any two vertices v and w of D, there is a chain from v to w. The digraph D_1 in Fig. 6.3 is not strongly connected because there is no chain from r to p (although there is one from p to r). A directed graph is *Eulerian* if there exists a closed directed path which includes every arc of D just once, and it is *Hamiltonian* if there exists a closed directed circuit which includes every vertex of D just once. It is known (Ghouila–Houri 1960) that if D is a strongly connected directed graph with n vertices, and both the out-degree and in-degree of each vertex are at least $\frac{1}{2}n$, then D is Hamiltonian. A connected directed graph D is Eulerian if and only if the in-degree and out-degree are equal for each vertex of D.

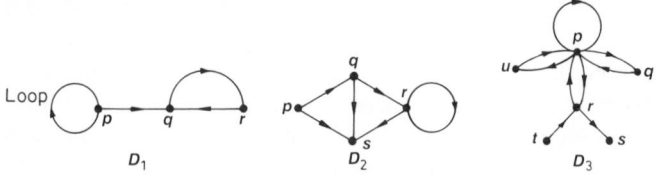

Fig. 6.3 Directed graphs

1.7 Some common networks

(a) *Linear and cyclic*

Linear connectivity corresponds to the simplest connected graph on $v_1,..., v_n$ in which vertices are joined only if adjacent in this list; in practice, to give additional flexibility, the first and last elements would be connected to make a cyclic configuration (see Fig. 7.1). Any processor P_j can directly access data from its adjacent neighbours P_{j-1} and P_{j+1}.

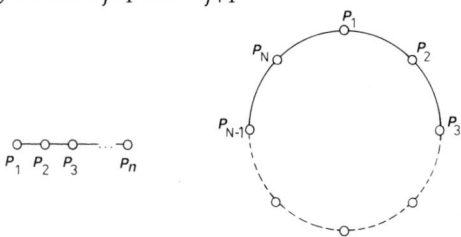

Fig. 7.1 Linear and cyclic connectivity of order n

The linear connectivity may be applied, for example, to the following calculation (arising from the solution of a differential equation).

given x_i $(i = 1,..., n)$
evaluate $y_i = x_{i-1} + x_{i+1} - 2x_i$ for $i = 2,..., n - 1$.

Method
assume $x_1,..., x_n$ are stored in $P_1,..., P_n$ respectively
compute $y_i = x_{i-1} + x_{i+1} - 2x_i$ for all $i = 2,..., n - 1$
             simultaneously.

(b) *Perfect shuffle*

The perfect-shuffle interconnection permits any route, but more importantly it is the basis of an arbitrary permutation, so routeing capacity can be high. It is most appropriate for sorting numbers using the bitonic sorting algorithm. Here we discuss the perfect shuffle and the associated topology; details of the bitonic sort implementation are given in Chapter 3.

Suppose that a pack of $n = 2^m$ playing cards, indexed as $C_0,..., C_{n-1}$ from the top downwards, is cut into two equal halves

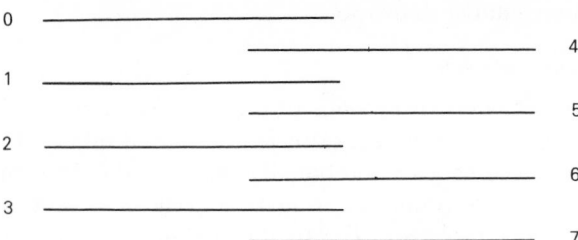

Fig. 7.2 Perfect shuffle for $n = 8$

consisting of the top $\frac{1}{2}n$ cards and bottom $\frac{1}{2}n$ cards respectively, and that these are then interleaved in the way shown in Fig. 7.2 for the case $n = 8$.

This process is known as the perfect shuffle, and is approximated by the 'riffle' used by experienced card players. What is of interest here is the resulting permutation π, where $\pi(i)$ is the index of the card that occupies the ith position after the shuffle $(i = 0,...,n-1)$:

$$\pi: \begin{bmatrix} 0 & 1 & 2 & 3 & 4 & \ldots & n-2 & n-1 \\ 0 & \frac{1}{2}n & 1 & \frac{1}{2}n+1 & 2 & \ldots & \frac{1}{2}n-1 & n-1 \end{bmatrix}$$

In cyclic notation for $n = 8$, this is $(0)(142)(356)(7)$. Here 0 is connected to itself: 1 is connected to 4 which is connected to 2 which, in turn, is connected back to 1; and so on. It is easy to see that $\pi^{-1}(x) = 2x$ (modulo $n-1$) for $x = 0,...,n-2$, with $\pi^{-1}(n-1) = n-1$. We can form a directed graph D_m on $n = 2^m$ vertices $0,...,n-1$ by connecting

(i) $\pi^{-1}(x)$ to x for $x = 0,...,n-1$,
(ii) $2x$ to $2x+1$ and $2x+1$ to $2x$ for $x = 0,...,\frac{1}{2}n-1$.

For $m = 3$ (i.e. $n = 8$) the connections are thus as shown in Fig. 7.3 and (as a plane graph, with nonintersecting arcs) in Fig. 7.4. For $m = 4$ (i.e. $n = 16$) the graph is displayed in Fig. 7.5.

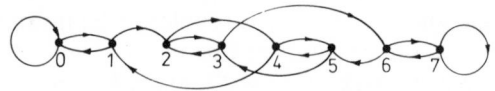

Fig. 7.3 Directed graph D_3

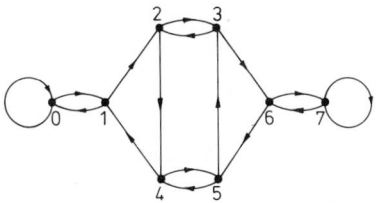

Fig. 7.4 Directed graph D_3

For parallel computation, processors $P_0,\ldots, P_{n-1}$ are connected as follows:

$$P_i \rightarrow P_{\pi^{-1}(i)} \quad (i = 0,\ldots, n-1),$$
$$P_0 \leftrightarrow P_1,\ldots, P_{n-2} \leftrightarrow P_{n-1}.$$

We next consider the properties of D_m. The graph is strongly connected: for $v \neq w$, there exists a chain from v to w. However, D_m does not have a Hamiltonian circuit, but is Eulerian because the out-degree is equal to the in-degree for every vertex (see Section 1.6). In practice this means that, through 'local' connectivity alone, the interconnection provides enough generality to

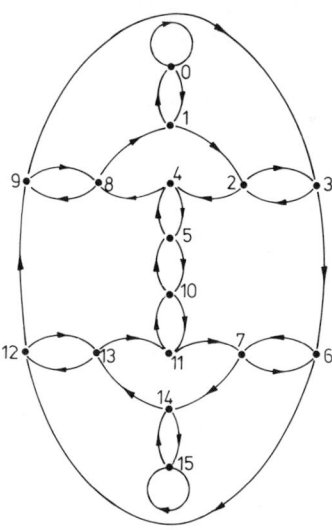

Fig. 7.5 Directed graph D_4

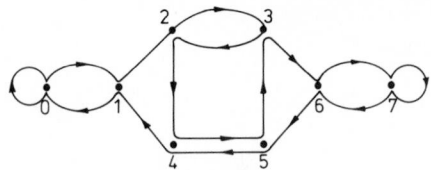

Fig. 7.6 Eulerian path for D_3

transmit a message from one processor to any of the others: for example, D_3 has the Eulerian dipath, shown in Fig. 7.6. In practice, because there are only four edges per vertex, the current design of Inmos transputers, with four links per transputer, may provide the basis for the construction of a transputer-based MIMD machine with perfect-shuffle interconnection.

(c) *Tree*

A *tree* is defined as a simple graph (no loops or multiple edges) with no closed circuits and with one vertex—the *root*—distinguished. Figure 7.7 represents the binary tree B_n of order n, with $1 + 2 + 2^2 + \cdots + 2^{n-1} = 2^n - 1$ vertices and $2^n - 2$ edges.

Binary trees are associated with algorithms for the solution of a number of fundamental problems such as sorting, searching, evaluation of algebraic expressions, library classification, and making decisions in games: these may often be conveniently implemented on a parallel processor with binary tree connectivity. In such a system, each processor at level k is directly

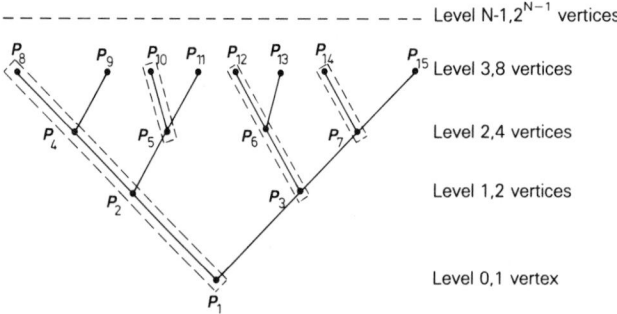

Fig. 7.7 Binary tree B_n of order n

connected to exactly zero or two processors at level $k + 1$; conversely, each processor at level $k + 1$ is directly connected to a single processor at level k, with the exception of the processor at level 0, which corresponds to the root node. Thus the degree of a vertex can be 1, 2, or 3. Any two processors are connected by a unique path.

In general, for an SIMD system with this connectivity, computations at the rth level are carried out simultaneously with all the processors operating under the control of the central processing unit. In the following example, it is assumed that the cost of passing data along any single link is the same for all links, which may not be true in practice.

Suppose we wish to evaluate the sum of $s = 2^{n-1}$ numbers.

Method   *Assume* $s = 8$ and $n = 4$ (from which the general case is evident).
           Store the numbers as $x_8, \ldots, x_{15}$ in $P_8, \ldots, P_{15}$.

Step 1   *add* x_8 and x_9 in P_4 and store the result as x_4
           (see Fig. 7.7).
           add x_{10} and x_{11} in P_5 and store the result as x_5, etc.

Step 2   *add* x_4 and x_5 in P_2 and store the result as x_2, etc.

Step 3   *add* x_2 and x_3 in P_1 and store the result as x_1.

Computations within each level are carried out simultaneously; and clearly, in exactly $n - 1 = \log_2 s$ steps, the sum appears in P_1.

Essentially the same algorithm can be implemented on machines with the 'condensed binary tree' processor connectivity shown in Fig. 7.8 (for the case $n = 4$).

The arrows indicate the address (processor) to which data are moved and added in successive steps: at Step 1, the number x_{2n+2} is added to x_{2n+1} and stored in P_{2n+1} (i.e. $x_{2n+1} \leftarrow x_{2n+1} + x_{2n+2}$), and so on. The connectivity is that of the binary tree in Fig. 7.7 if processors on the same North-West to South-East line are

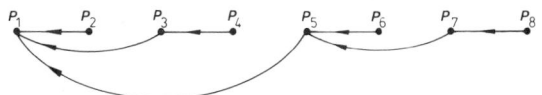

Fig. 7.8   Condensed binary tree of order 4

identified: P_8, P_4, P_2, P_1 are all equal to P_1; P_9 is P_2; P_{10} and P_5 are equal to P_3; and so on (dotted lines in Fig. 7.7 enclose processors that are identified).

The condensed binary tree C_n of order n is defined recursively as follows. C_1 is the simple tree with one vertex, and

$$C_{k+1} = [C_1, \dots, C_k] \quad (k \geq 1),$$

where $[C_1, \dots, C_k]$ is the tree formed by joining each of the trees $C_1, \dots, C_k$ by its root to a new vertex which is the root of C_{k+1}. Thus C_n has 2^{n-1} vertices.

(d) *Lattice mesh*

The graph of a multiprocessor with lattice (square grid) architecture consists of n^2 nodes, each connected to four neighbours: the case $n = 4$ is shown in Fig. 7.9; for clarity, the 'horizontal wrap-around' connections are *not* shown.

If the processors are numbered $P(i, j)$ ($i, j \in \{1, \dots, n\}$), then $P(i, j)$ is connected to

$$P(i, i-1), \quad P(i, j+1), \quad P(i-1, j), \quad P(i+1, j). \quad (7.1)$$

Usually the processors at the edges are connected cyclically: $P(1, 1)$ to $P(1, n)$ and $P(n, 1)$; $P(1, 2)$ to $P(n, 2)$; and so on. This means that $i \pm 1$ and $j \pm 1$ are taken modulo n in (7.1), and the lattice is most correctly thought of as $\mathbb{Z}_n \times \mathbb{Z}_n$ and lying on a torus rather than on the plane. In Fig. 7.9 *half these additional connections* are indicated with dotted lines. This provides extra flexibility and reduces the diameter of the graph from $2n - 2$ to $2\lfloor \frac{1}{2}n \rfloor$, where $\lfloor \bullet \rfloor$ denotes the integer-part function.

This graph has a much higher degree of symmetry than those described earlier, and the implications of this will be discussed in the next section. It is easy to see that it has enough connections to embed the cyclic graph of order n^2 with direct links.

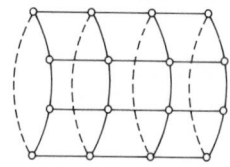

Fig. 7.9 Two-dimensional lattice of order 4

In addition to local connectivity (displayed in Fig. 7.9), for an SIMD system of this type, the processors may also communicate via global buses providing, for example, row and column highways—fast transmission lines for passing messages along rows and columns. A number of examples involving use of the lattice are given throughout this book, particularly for numerical methods that employ n-dimensional lattices with $n \geq 2$.

(e) *Hypercube*

The 2^n vertices and $2^{n-1}n$ edges of an n-dimensional cube form the nodes and edges of the *hypercube graph* H_n: if the nodes are represented by binary numbers $\varepsilon_1, \ldots, \varepsilon_n \in \{0, 1\}$, then two nodes are joined if and only if their representations differ in exactly one position. The cases $n = 1$, $n = 2$ (square), and $n = 3$ (cube) are shown in Fig. 7.10.

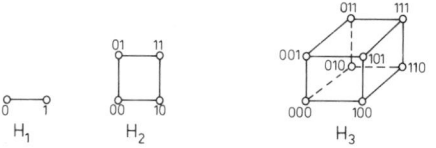

Fig. 7.10 Hypercubes H_1, H_2, H_3

Clearly, H_{n+1} can be constructed by joining the corresponding nodes of two copies of H_n: the cases $n = 3, 4$ are displayed in Fig. 7.11. Each node is joined directly to n other nodes. The diameter of H_n is n, so that the distance between any two nodes is relatively small compared with the size of the graph. This is significant because, in a parallel computer with this topology, the cost of passing messages between two elements is also likely to be small. On the other hand, realization does require each element to have n connections.

Parallel computers with hypercube topology of up to 10 dimensions are currently in use, and available commercially as integral parts of MIMD systems: higher dimensions will no doubt appear soon. The properties of hypercubes are being investigated with respect to a variety of applications, and particularly the design of numerical algorithms. For further reading, see the paper by Saad and Schultz (1986) on the cost of data communication, which shows that hypercube topology can be very efficient in performing certain data exchange operations.

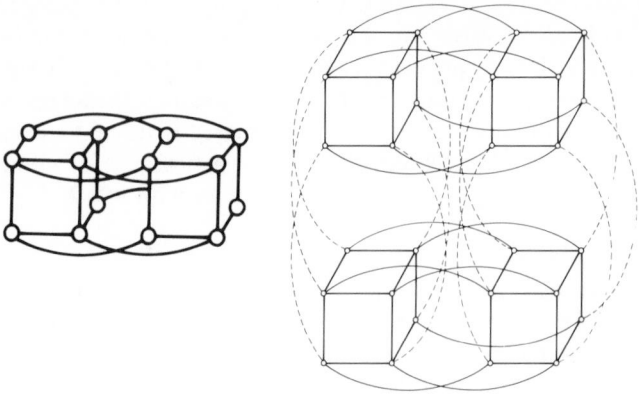

Fig. 7.11   Forming H_4 from H_3, and H_5 from H_4

1.8   Symmetry, homogeneity, and embedding

The presence of symmetry and homogeneity in the commercially
available machines is no mystery: these properties impart con-
ceptual simplicity and flexibility in use, and decrease the need for
programs to take account of different types of processing
elements within a given network. The necessity for embedding
arises when a programmer wishes to implement an algorithm A
for which it is clear that a certain network G is most appropriate,
but only another network H is available. The Bitonic sort
algorithm, for example, ideally requires the perfect-shuffle interconnection, but the programmer may nevertheless wish to
implement it on machines based on the lattice or the hypercube.
The problem can be solved by utilizing part of H as a model of
G, in other words: *embedding G in H*. Thus the perfect-shuffle
directed graph D_3 (Fig. 7.3), with arrows and loops removed and
multiple edges replaced by single edges, becomes the graph P_3 in
Fig. 8.1, which can be embedded in the 3×3 and 4×4 lattices L_3
and L_4 in the indicated ways.

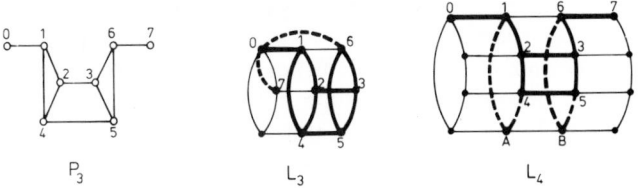

Fig. 8.1 Embedding P_3 in L_3 and L_4

The embeddings are not *direct*: the edge 67 in P_3 is represented by the pair 60–07 in L_3, and the edge 41 in P_3 is represented by 4A–A1 in L_4 and 56 by 5B–B6 (shown by broken lines in Fig. 8.1). In order to minimize data movement costs, such indirect links should be short; depending on the particular machine involved, it may or may not be important to avoid using one of the processing elements (such as 0 in L_3) both as an active processor and also as an intermediate stage in data movement: in this respect, the embedding L_4 might be preferable, since here A and B have only the second function. It might also be desirable in some contexts to avoid using any one edge of H in two functions, i.e. both as a direct link between one pair of elements and as part of an indirect link between other pairs, which would happen if the edge 67 in P_3 were represented by 63–37 in the embedding in L_3.

An important case is when a three-dimensional situation (for example in a finite-element problem), where perhaps the hypercube or some kind of 'hyperlattice' would be ideal, must be dealt with using an essentially two-dimensional network such as the lattice. Even with a reconfigurable array of processors, this difficulty may arise, either because of the cost of reconfiguration or because of a restriction on the maximum degree of vertices in any available configuration. One of the skills in parallel computing is to find a convenient embedding, involving a minimum use of long links, of one network in another. We shall discuss briefly the degree of homogeneity and symmetry possessed by these graphs, as well as the extent to which they can be embedded in lattices and hypercubes.

(a) Linear and cyclic

The linear connection J_n of order n has minimal symmetry: it can merely be reversed. The cyclic connection O_n has more—it can be rotated as well. Furthermore O_n is homogeneous, and any given pair of adjacent points will map to any other such pair, and indeed any pair of vertices at distance d will map to any other pair at this distance, under some automorphism. Although the cycle is very homogeneous, it is not very symmetric: the number of automorphisms of O_n is only $2n$.

Each cycle O_n can be embedded in a sufficiently large lattice or hypercube. One indirect link of length 2 may be necessary if n is odd, because all cycles in a hypercube are of even length; the same is true of all cycles in J_m if m is even, and of all cycles of length less than m if J_m is odd.

(b) Hypercube and lattice

The hypercube H_n of order n is very symmetric—it has $2^n(n-1)!$ automorphisms. It is also very homogeneous: for example, any two (or three) points, with mutual distances given, will map to any other two (or three) points with the same mutual distances. This, however, is not true for the lattice L_n of order n, since no automorphism maps ⌐o—o to o—o—o. On the other hand, any square ⌐⌐ can be mapped to any other square, in both H_n and L_n.

(c) Binary tree and condensed binary tree

The binary tree B_n of order n is not homogeneous, because no automorphism will take a vertex to one at a different level, but it has a high degree of symmetry: the total number of automorphisms is $2^{2^{n-1}-1}$. Binary trees of order 2, 3, 4, and 5 can be embedded with direct links in lattices of order 2, 3, 5, and 7 (see Fig. 8.2), but a binary tree of order 6 (or larger) cannot be embedded in a lattice of any order. This is because the binary tree of order 6 has all its 63 points at a distance at most 5 from one particular point, the root, while a lattice graph contains at most 61 such points.

It can be shown that the condensed binary trees C_2, C_3, C_4, and C_5 can be embedded, using direct links, in L_2, L_2, L_3, and L_4 respectively.

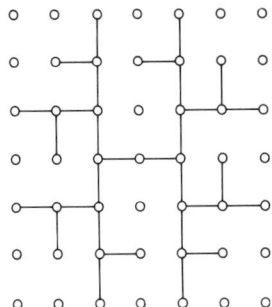

Fig. 8.2   Direct embedding of B_5 in L_7

1.9   Pipelining and vector processing

In this section, we revert to the case of a single CPU (and therefore to what is essentially a serial machine), but with a limited amount of parallelism incorporated. Certain parts of the CPU which are responsible for separate functions (fetching operands, executing arithmetic operations, outputting) can be instructed to operate simultaneously.

As a simple example, suppose that we wish to add two vectors $(a_i : i = 1,..., n)$ and $(b_i : i = 1,..., n)$, and that the addition of two numbers a and b is broken into three subtasks, each carried out by a separate subunit of the CPU:

fetch a ; fetch b and add to a ; output $a + b$.

(The situation is more complicated in reality, because each of these simple operations itself consists of two or more subtasks.) Then we can arrange for all three subunits to be in operation simultaneously throughout most of the calculation.

Thus, in clock period 1, a_1 is fetched; in clock period 2, a_2 is fetched and simultaneously b_1 is fetched and added to a_1; in clock period 3, a_3 is fetched and simultaneously b_2 is fetched and added to a_2 while $a_1 + b_1$ appears in the output; a similar process continues until the calculation is complete. We can indicate this by showing, in Table 9.1, how each of the three subunits is occupied in each clock period.

Clearly the time is reduced to approximately one-third of that required otherwise, if all subtasks are assumed to take the same

Table 9.1 Pipelining of three operands

Clock period:	1	2	3	4	... n	$n+1$	$n+2$
Fetch:	a_1	a_2	a_3	a_4	... a_n	—	
Fetch and add:	—	a_1+b_1	a_2+b_2	a_3+b_3	... $a_{n-1}+b_{n-1}$	$a_n+\ldots$	
Output:	—	—	a_1+b_1	a_2+b_2	... $a_{n-2}+b_{n-2}$	$a_{n-1}+\ldots$	

time. This technique is described as pipelining. It had been envisaged from the earliest days of computing and was first incorporated in machines such as EDSAC 1 at Cambridge, and ACE at the National Physical Laboratory.

There is a more extensive use of pipelining in the modern supercomputer and other large mainframe machines or *vector processors,* where the arithmetic unit of the CPU may be regarded as being made up of separate subunits which can be instructed to operate simultaneously and feed the results into other subunits. This is known as *chaining.* Thus, for example, if we wish to calculate $(a_i + b_i)c_i$ $(i = 1,..., n)$, where (a_i), (b_i), and (c_i) are three n-vectors, then (ignoring the other subtasks such as fetch and output) we can organize the calculations as shown in Table 9.2.

We can do a simple analysis as follows. Suppose that m is the number of subtasks required to perform some operation, and that τ is the time for each subtask (for simplicity it is assumed that each takes the same time); then, to obtain one result takes $m\tau$ units of time, and the rest appear in succession, so that n results will appear in a single vector operation in $(n - 1)\tau + m\tau$ units of time. If $\sigma = (m - 1)\tau$, then the time taken is $n\tau + \sigma$. On a serial machine without pipelining, the time taken is nmt, where

Table 9.2 Pipelining of addition and multiplication

Clock period:	1	2	3	...
Add:	a_1+b_1	a_2+b_2	a_3+b_3	...
Multiply:	—	$(a_1+b_1)c_1$	$(a_2+b_2)c_2$	...

t is the time for each subtask. Thus vector operation is clearly better than scalar operation provided that

$$nmt > n\tau + \sigma, \text{ i.e. } n > \sigma/(mt - \tau), \text{ with } mt > \tau.$$

In general, better performance can be expected when the machine is used for lengthy repetitive calculations rather than conditional operations.

Examples of vector processors include the CDC Star 100, the Floating Point series (AB-120B, AP-164, AP-5000), the Cray series (Cray-1, Cray-XMP, Cray-2, Cray-3), and the Cyber 205. A number of them have a hybrid design incorporating multiple processors, each of which utilizes vectorization. In the Cray series, for example, Cray-2 can have one, two or four processors operating simultaneously, and Cray-3 can have up to sixteen.

The Cray-1 with a single processor makes extensive use of vector processing with 12 independent pipelines to provide a full range of scalar and vector operations. Of these, there are three arithmetic pipeline units for addition/subtraction, multiplication, and forming the reciprocal. Table 9.3 shows the number of subtasks for these operations. The bipolar memory is interleaved into blocks of 16 with the total size of 1M 64-bit words (8 Mbytes). The cycle time of the memory is 50 nsec, but, by phasing the operation of each memory block, it is possible to transfer a 64-bit word to the internal registers in 12.5 nsec, which is the cycle time of the processor.

Suppose we perform the vector operations $x = y * z$, where $*$ indicates element-by-element multiplication, followed by $a = b + x$. The first element of x will appear after 9 clock cycles (including one clock cycle each for moving operands to and from the register to the multiplication unit), at which time the calculation of $b + x$ can start, provided that both b and the adder are free.

Table 9.3 Number of subtasks for floating-point arithmetic on the Cray-1

Floating-point operation	Number of subtasks
Addition/subtraction:	6
Multiplication:	7
Reciprocal-formation:	14

Chaining will then allow the first element of a to appear after a further six clock cycles, i.e. a total of $15 \times 12.5 = 187\frac{1}{2}$ nsec. In vector arithmetic there is an initial start-up time, but subsequently each element of the resultant vector will appear in a minimum time of 12.5 nsec. For a detailed description of the Cray series, the reader is referred to Iliffe (1982).

In order to achieve optimal speed, it may be necessary to perform floating-point addition, floating-point multiplication, and fixed-point arithmetic, as well as to access operands to and from the main memory, all in parallel. However, there are often restrictions on inputs to the arithmetic units which make it difficult to exploit the arithmetic speed of the machine fully (see Modi and Rollett (1986) for a discussion on the design of the AP-120B).

1.10 Algorithm performance measurement

Two theoretical indices have been used for measuring the performance of a parallel algorithm:

$$\text{Speed-up:} \quad S_p = T_1/T_p,$$

where T_p is the time required for the calculation on a p-processor machine, and

$$\text{Efficiency:} \quad E_p = S_p/p.$$

The theoretical maximum value of S_p appears to be p (and of E_p to be 1), attained when the algorithm is fully parallel and the calculation is distributed equally among the p processors (processing elements). This would require all the processors to be operable simultaneously, each processor to be capable of performing all the arithmetic operations ($+$, $-$, $*$, $\div$), and the incurring of no memory data-movement penalties. The time may be thought of as measured in clock periods. When actual performance (possibly of different algorithms for a given problem) on particular machines (serial and parallel) is compared, one may study, for example, the number of megaflops or megaflops per second, or simply the total time taken for the calculation.

There are, however, good reasons for regarding all such theoretical indices as of only limited value in estimating the efficiency of a given algorithm on a given machine, since not only may the theoretical maximum be unapproachable in practice, but in certain circumstances this may be less important than would at first appear.

Consider a typical problem to be solved using a parallel computer, which involves a sequence of numbers $n_1, \ldots, n_k$ stored in the respective processors $P_1, \ldots, P_k$ and continually updated in the course of computation. Some relevant considerations are as follows.

1. The value of k may be smaller than the machine size (number of processing elements), and there may be no convenient programming techniques to split the data into smaller units for distribution among all the processors, so that non-utilization of some processors results.

2. Often k can be varied and should be made as large as possible, for example to reduce the mesh size in a discrete approximation to a continuous process. Associated with the process is a graph on $\{1, \ldots, k\}$ in which i is connected to j when the value of n_j contributes to the updating of n_i. It may be that every suitable embedding of this graph into the routeing network (with short paths—see Sections 1.7 and 1.8) inevitably bypasses a proportion of the processors, so that k cannot approach the theoretical maximum of the machine size. For example, if a problem involves the values of some physical quantity at certain grid points on the surface of a sphere, and implementation is carried out using a routeing network with lattice connectivity, then a significant percentage of the processors may be idle at any given moment. In such a situation, additional computing time will be needed to reach a given accuracy of solution.

3. Where, as in (2) above, the routeing network is not fully compatible with the natural graph of the problem, and embedding therefore involves indirect paths, extra clock periods may be used in order to move data between processors, for example, from P_1 to P_4 via P_2 and P_3, and a substantial overhead may be incurred.

4. In the implementation of conditional instructions, parallelism may be lost in more than one way. First, at the instant when a conditional test on a single scalar value is made, only a single processor may be in use. Second, a branch instruction may

require different operations to be performed on certain of the variables $n_1,\ldots,n_k$, but—in an SIMD system—these operations cannot be performed simultaneously, and thus at any given moment a subset of the processors will be idle.

As an example of a parallel algorithm in which an apparently severe underutilization of the processors is not of primary importance, we mention the process of finding the maximum of 2^p numbers on a 2^p-processor machine, by the method of recursive doubling (see Chapter 2). At successive stages only 2^{p-1}, 2^{p-2},..., $2^{p-p}=1$ processors are in use, and the overall processor utilization is only approximately $1/p$. However, in spite of certain data-movement costs, on a bit-oriented machine such as the DAP the method is highly efficient, taking only 50 μsec with $2^p = 4096$, less than it would take to subtract two numbers on a single processor. This illustrates that it is not necessarily disastrous to disable some of the processing elements. In fact, in a bit-organized system, suitable algorithms based on only a proportion of the available processing elements often show the best performance.

Most algorithms incur two basic cost components:

(i) arithmetic, under which we subsume logical operations, which are not distinguishable if the processing elements operate at bit level;

(ii) data movement.

For a realistic assessment of performance, these two factors must be taken into account. In a survey of the relative performance of a whole spectrum of parallel computers, Hockney (1986) considers each of the parameters—vector length, work per memory reference, and granularity—as a function of the number of processors and the complexity of the individual processors, and gives a final measure of performance for a number of computers, including Cray-XMP and the FPS-5000 series.

1.11 Interplay: parallel computers v. parallel algorithms v. problem size

The choice of algorithm strongly depends on the various parameters affecting the execution time. The timing characteristics of parallel computers are different from those of serial computers, and the choice of an optimal algorithm may differ accordingly. This section is based on Parkinson's (1982a) comparison of the way algorithm performance on serial, parallel, and vector processors varies with the size of the problem (such as the mesh size for a finite-element analysis) and the number of processing elements available on a given parallel system.

Theorists often assume an abstract parallel machine of sufficient size to ensure that, when an operation is to be performed on N data items, this can be done in one parallel step. In practice, the required number of processors may not be available. Similarly, the vector processor may accept vectors of no more than some fixed length. Taking these limitations into account, and including the case of a serial machine for comparison, we consider first the total time (T, with appropriate suffix) needed to perform the same operation on N data items, in terms of the time (t, with appropriate suffix) for a single operation (for a subtask in the case of a vector processor) and other characteristics of the machine.

(a) Serial processor: $\quad T_s = Nt_s$
(b) Parallel processor: $\quad T_p = \lceil N/M_p \rceil t_p$

Here M_p is the number of available processing elements: at most M_p data items can be processed simultaneously, and the data must therefore be partitioned into groups of M_p (or fewer) items; each group can be processed simultaneously, and there are $\lceil N/M_p \rceil$ groups, where $\lceil x \rceil$ denotes the smallest integer $\geq x$.

(c) Vector processor: $\quad T_v = Nt_v + \lceil N/M_v \rceil s_v$

Here s_v is the start-up time for a vector operation and M_v the maximum length of vector that can be processed for each start: the data must be partitioned into groups (vectors) of M_v or fewer items; each group is processed in time $nt_v + s_v$, where n is the number of items in the group, and there are $\lceil N/M_v \rceil$ groups.

We see that the timing, as a function of N, depends on three parameters (t_v, s_v, and M_v) in the case of a vector processor, and on only one parameter (t_s) or two (t_p and M_p) in the cases of serial and parallel processors. The vector processor thus has hybrid characteristics.

If we let $R_i = T_i/t_i$ ($i = $ s, v, p), in order to remove the effects of the relative speeds, then we obtain

$$R_s = N,$$
$$R_p = \lceil N/M_p \rceil,$$
$$R_v = N + \lceil N/M_v \rceil q_v \quad (q_v = s_v/t_v).$$

It is useful to compare the graphs of R_i against N for the three cases (Fig. 11.1(i)).

For the real systems: distributed array processor (DAP) (with $M_p = 4096$) and Cyber 205 (with $M_v = \infty$ and $q_v = 40$), graphs of $\log_2 R_i$ against $\log_2 N$ are drawn on logarithmic scale for ease of representation (Fig. 11.1(ii)). For the vector processor, the curve differs from the serial machine for small values of N. However, the DAP curve has two distinct segments, with a linking section; performance measurement along each segment is identified.

(i) The α section of Fig. 11.1(ii) shows that performance of an algorithm on the DAP in this range does not depend on the dimension of the problem. New formulations for performance measurement are therefore needed in this range. This is reflected in the design of algorithms where it is assumed that enough processors are available to match the dimension of the problem; for example, the Gauss–Jordan method for solving a linear system of equations shows a high operation count on a serial machine; nevertheless, when it is mapped onto an array of processors, a high performance is obtained (see Chapter 4, Section 4.2).

(ii) The β section of Fig. 11.1(ii) refers to a problem size N that is larger than the number of processors M_p but is only a small multiple of M_p. The data cannot be fully accommodated by assigning each data item to a separate processing element; therefore the data are partitioned into $\lceil N/M_p \rceil$ groups each consisting of M_p (or fewer) items, and techniques such as those of 'slicing' and 'crinkling' (see Chapter 2) are used to retain as much parallelism as possible. For example: if M_p is 16, and we have 40 data items, then three groups are required.

(iii) The γ section of Fig. 11.1(ii) indicates a problem size which is very large compared to the number of processors available and performance characteristics are similar to those for serial machines.

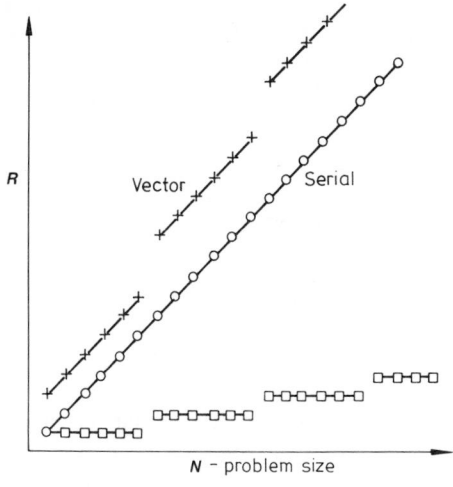

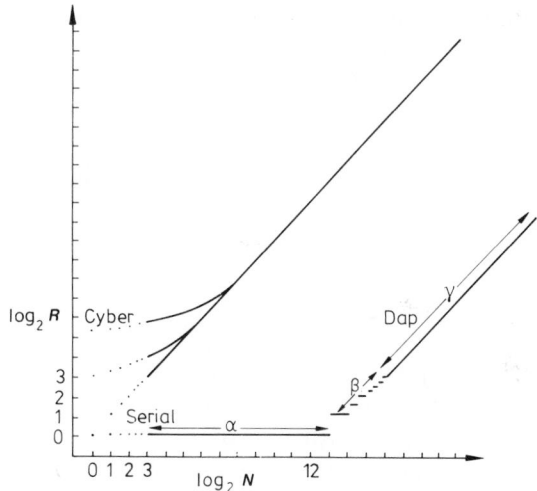

Fig. 11.1 Performance dependence on problem size

We take an example of multiplication of two $n \times n$ matrices, assume that at least n^2 processors are available, and consider the timings on a serial computer and the DAP (case (i) above). For any serial method, n^3 multiplications are required. For an $n \times n$ DAP with $n \le 64$, n^2 multiplications can be done simultaneously

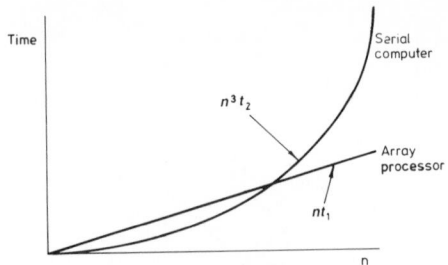

Fig. 11.2 Variation of matrix multiplication time on the DAP and a serial machine

(see Chapter 2), so that time varies linearly with n. Figure 11.2 shows the time dependencies for the two methods.

The coefficients t_1 and t_2 represent the inner-loop calculations (+, *, indexing, etc.) for the serial method and the DAP, which take time independent of n, provided that $n \leq 64$.

One version of the parallel method—the outer-product algorithm—proceeds as follows (also see Chapter 2). In forming the matrix product $C = AB$ with elements

$$c_{ij} = \sum_{k=1}^{n} a_{ik} b_{kj} \quad (1 \leq i \leq n, \, i \leq j \leq n),$$

there are n^3 products $a_{ik} b_{kj}$ to be calculated. These are calculated as:

$$C = AB = [\mathrm{c}^n(A_{\bullet 1}) * \mathrm{r}^n(B_{1 \bullet})] + \cdots + [\mathrm{c}^n(A_{\bullet n}) * \mathrm{r}^n(B_{n \bullet})],$$

where $\mathrm{c}^n(A_{\bullet k})$ denotes the $n \times n$ matrix consisting of the kth column of A, repeated n times, and $\mathrm{r}^n(B_{k \bullet})$ denotes the $n \times n$ matrix consisting of the kth row of B, repeated n times. The calculation is therefore reduced to n applications of termwise multiplication *.

The inner-loop calculation: $\mathrm{c}^n(A_{\bullet k}) * \mathrm{r}^n(A_{k \bullet})$ consists of (i) addition and multiplication, and (ii) extracting a selected column (or row) of data from a processor and propagating it by rows (or columns). Provided that $n \leq 64$, these operations take a time *independent* of n. Thus, in this case, matrix multiplication takes a time of nt_2, where t_2 is the inner-loop time. The increase in speed over the serial machine is $n^2 t_1/t_2$, which indicates the dependence on the size of the application; in particular, the maximum value is attained when the size of the matrices matches the number of processors.

The above example illustrates a number of important points: (i) the algorithms for parallel computation are different from those for serial computation, and, in order to measure real performance on parallel systems, recoding must be permitted; (ii) the performance can be a function of problem size—if the problem size is not chosen to be optimal then one is not measuring the true performance; (iii) even for similar architectures, for example the SIMD systems—DAP with 64×64 processors and MPP with 128×128 processors—the performance can vary significantly. With regard to the last point, the crucial factor which makes the DAP matrix multiplication linear in matrix size is the existence of row and column highways which permit fast transmission of data. The MPP does not possess such a network, and the cost of inner-loop calculation becomes proportional to n. Thus the characteristics for matrix multiplication on MPP are qualitatively different from those on DAP and the serial machine.

2 Parallel techniques and algorithms

2.1 Introduction

Compared with serial systems, parallel systems permit more freedom of expression in problem analysis and programming. A foundation in the skills of thinking in parallel is basic to the understanding of such systems. This chapter is devoted to techniques and examples illustrating the design of parallel algorithms.

The following general principles may provide some useful guidelines.

(a) The most obvious approach is to examine a serial method and convert it into a procedure that operates on composite mathematical objects such as vectors and matrices, so that many data are processed simultaneously. However, the latest and most efficient serial method is not always best suited to such adaptation: often an earlier, less efficient, serial method already possesses a high degree of parallelism, and so is much more adaptable to parallel computation.

(b) It may turn out that an iterative algorithm equivalent to a certain direct (i.e. noniterative) method possesses a higher degree of parallelism (more independent computations) and can be organized systematically on a parallel machine. In this case, it is important to prove the stability of the method, and to ensure that the rate of convergence is acceptable. It should, of course, be shown that this is a cost-effective procedure when compared with the direct method.

(c) The computation may be broken down into smaller units and distributed among the processors. Two techniques based on this principle are discussed in Sections 2.5 and 2.6.

We discuss the above ideas in connection with the following

topics: (i) elementary parallel operations, (ii) matrix multiplication, (iii) parallel evaluation of arithmetic expressions, (iv) recursive doubling, (v) cyclic odd–even reduction, (vi) bit-level algorithms, (vii) bit-level algorithms for numerical functions, (viii) slicing and crinkling, and finally (ix) parallel data transforms—a calculus of data organization.

In each case the emphasis is on the identification of parallelism, and its systematic exploitation. No attempt has been made to discuss the details of implementation on any particular type of machine, though suitability for SIMD or MIMD implementation (see Chapter 1, Section 1.2) is indicated where appropriate.

We begin with a brief discussion of a number of elementary parallel operations, which will also serve to establish some notation for later use.

2.2 Elementary parallel operations

The following operations are likely to be available on most parallel computers, and the attendant definitions are fundamental to the ensuing discussion. The natural assumption here is that each processor is capable of performing arithmetic on its *local* data, which are assumed to be located in the required position.

(i) *Arithmetic operations:* $+$, $-$, $*$, $\div$ denote termwise addition, subtraction, multiplication, and division respectively on objects such as matrices and vectors; for example, for vectors v and w:

$$v + w = (v_1,\ldots, v_n) + (w_1,\ldots, w_n) = (v_1 + w_1,\ldots, v_n + w_n),$$

and

$$v * w = (v_1,\ldots, v_n) * (w_1,\ldots, w_n) = (v_1 * w_1,\ldots, v_n * w_n).$$

(We note that the symbols for the termwise operators $+$, $-$, $*$, $\div$ are not distinguished from those for the corresponding scalar operations.)

(ii) *Row and column selection operations:* $A_{i\bullet}$ and $A_{\bullet j}$ denote the ith row and jth column of matrix A. For example, if $A = \begin{bmatrix} 1 & 2 \\ 3 & 4 \end{bmatrix}$, then $A_{1\bullet} = [1 \quad 2]$ and $A_{\bullet 1} = \begin{bmatrix} 1 \\ 3 \end{bmatrix}$.

(iii) *Row and column operations:* $\sum^r A$ and $\sum^c A$ denote the row and the column vector obtained by summing the rows and columns of A, respectively; for example, for A as in (ii) above $\sum^r A = [4, 6]$ and $\sum^c A = \begin{bmatrix} 3 \\ 7 \end{bmatrix}$.

(iv) *Matrix formation operations:* $r(\boldsymbol{v}_1^T, \ldots, \boldsymbol{v}_n^T)$ and $c(\boldsymbol{v}_1, \ldots, \boldsymbol{v}_n)$ denote those matrices whose rows or columns respectively are the vectors $\boldsymbol{v}_1^T, \ldots, \boldsymbol{v}_n^T$ and $\boldsymbol{v}_1, \ldots, \boldsymbol{v}_n$, where T denotes transpose.

(v) *Maximum and minimum operations:* $\max A$ and $\min A$ denote the maximum and minimum elements of a vector or matrix A.

(vi) *Logical operations:* AND and OR denote the termwise Boolean $\wedge$ and $\vee$ respectively on objects such as logical matrices and vectors; for example, for logical vectors $\boldsymbol{m}$ and $\boldsymbol{p}$:

$$\boldsymbol{m} \text{ AND } \boldsymbol{p} = (m_1, \ldots, m_n) \text{ AND } (p_1, \ldots, p_n)$$
$$= (m_1 \text{ AND } p_1, \ldots, m_n \text{ AND } p_n).$$

2.3 Matrix multiplication

Let A and B be matrices of size $m \times n$ and $n \times p$ respectively. In forming the matrix product $C = AB$ with elements

$$c_{ij} = \sum_{k=1}^{n} a_{ik} b_{kj} \quad (1 \leqslant i \leqslant m, \, 1 \leqslant j \leqslant p), \tag{3.1}$$

there are mnp products $a_{ik} b_{kj}$ to be calculated.

We will consider various strategies for forming this product on a parallel computer with N processors for each of the following cases:

(a) $N > mnp$,
(b) $N \geqslant \max(mn, np, mp)$,
(c) $N < \max(mn, np, mp)$.

(a) $N > mnp$

The matrix $C = AB$ has mp entries; each is the sum of products of n pairs of numbers; and the total number of scalar multiplications is mnp. If $N > mnp$, then all multiplications can be performed with a single application of $*$, by multiple positioning of the data entries, as

$$
m \left\uparrow\downarrow \begin{bmatrix} A & A & \cdots & A \end{bmatrix} * \begin{bmatrix} B_{1\cdot}^{\mathrm{T}} & B_{2\cdot}^{\mathrm{T}} & \cdots & B_{p\cdot}^{\mathrm{T}} \\ \vdots & \vdots & & \vdots \\ B_{1\cdot}^{\mathrm{T}} & B_{2\cdot}^{\mathrm{T}} & \cdots & B_{p\cdot}^{\mathrm{T}} \end{bmatrix} \right\uparrow\downarrow m \,,
$$

$$
\longleftarrow np \longrightarrow \quad \longleftarrow np \longrightarrow
$$

where $B_{i\cdot}^{\mathrm{T}}$ denotes the ith row of B^{T}.

The p successive copies of A are placed in adjacent positions, and the columns of B are placed horizontally and repeated m times, as indicated above. For instance, for a 2×2 case, we have

$$
\begin{bmatrix} a_{11} & a_{12} \\ a_{21} & a_{22} \end{bmatrix} \begin{bmatrix} b_{11} & b_{12} \\ b_{21} & b_{22} \end{bmatrix}
$$

$$
= \begin{bmatrix} a_{11} & a_{11} \\ a_{21} & a_{21} \end{bmatrix} * \begin{bmatrix} b_{11} & b_{12} \\ b_{11} & b_{12} \end{bmatrix} + \begin{bmatrix} a_{12} & a_{12} \\ a_{22} & a_{22} \end{bmatrix} * \begin{bmatrix} b_{21} & b_{22} \\ b_{21} & b_{22} \end{bmatrix}.
$$

The products $a_{ik}b_{kj}$ are the elements of

$$
\begin{bmatrix} a_{11} & a_{11} & a_{12} & a_{12} \\ a_{21} & a_{21} & a_{22} & a_{22} \end{bmatrix} * \begin{bmatrix} b_{11} & b_{12} & b_{21} & b_{22} \\ b_{11} & b_{12} & b_{21} & b_{22} \end{bmatrix},
$$

which is more conveniently arranged as

$$
\begin{bmatrix} a_{11} & a_{12} & a_{11} & a_{12} \\ a_{21} & a_{22} & a_{21} & a_{22} \end{bmatrix} * \begin{bmatrix} b_{11} & b_{21} & b_{12} & b_{22} \\ b_{11} & b_{21} & b_{12} & b_{22} \end{bmatrix}.
$$

Various additions must be made, and the sums assigned to the correct positions in C. In general there are mp results in the result matrix, and each of the entries consists of the addition of n numbers (which takes $\lceil \log_2 n \rceil$ steps using recursive doubling; see Section 2.5). We can still obtain some additional parallelism by employing a similar strategy if N is sufficiently large, even if less than mnp.

(b) $N \geq \max(mn, np, mp)$

On a parallel machine with N processors, where $N \geq \max(mn, np, mp)$, it may be possible to perform mp, mn, or np of these multiplications simultaneously (in parallel) by means of the single operation $*$, of termwise multiplication. This will be done respectively n, p, or m times. Then (i) various additions must be made, and (ii) the sums assigned to the correct positions in C. If (i) and (ii) are of small cost compared with the operation $*$, then the overall cost will be minimized by minimizing the number of applications of $*$. The following strategies realize this objective. (In practice, if $N \geqslant M^2$, where $M \geqslant \max(m, n, p)$, it may be convenient to enlarge all matrices to $M \times M$, by introducing zeros in the additional positions.)

We now consider the three cases $\min(m, n, p) = n$, $\min(m, n, p) = m$, and $\min(m, n, p) = p$, the resulting algorithms being referred to as the *outer-product algorithm*, the *middle-product algorithm by rows*, and the *middle-product algorithm by columns* respectively.

Figure 3.1 shows a pictorial view of what is happening. Consider a cuboid of dimension $m \times n \times p$, where, in case (i), an

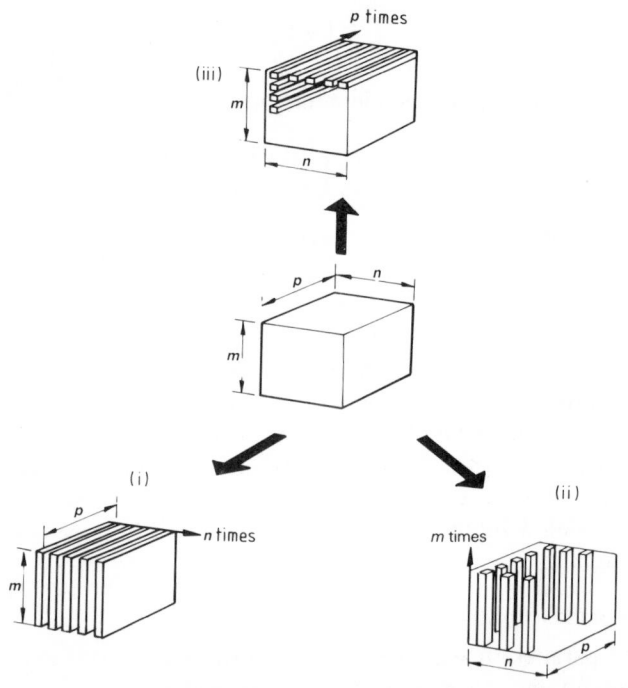

Parallel matrix multiplication: (i) outer-product algorithm, (ii) middle-product algorithm by rows, (iii) middle-product algorithm by columns

$m \times p$ array is formed in parallel at each step, and in n such steps the result is formed, as in Fig. 3.1(i). In case (ii), each column of the resulting matrix is formed in parallel at each step, and in m such steps the result is formed, as in Fig. 3.1(ii). Similarly, in case (iii), each row of the resulting matrix is formed in parallel at each step, and in p such steps the result is formed, as in Fig. 3.1(iii). Each case is considered in some detail below.

Case 1 $\min(m, n, p) = n$. In view of (3.1), we have $C = \sum_{k=1}^{n} C_k$, where $(C_k)_{ij} = a_{ik}b_{kj}$. We see that

$$
C_k = m \begin{array}{c} \uparrow \\ \\ \downarrow \end{array} \begin{bmatrix} a_{1k} & \cdots & a_{1k} \\ \vdots & & \vdots \\ a_{mk} & \cdots & a_{mk} \end{bmatrix} * \begin{bmatrix} b_{k1} & \cdots & b_{kp} \\ \vdots & & \vdots \\ b_{k1} & \cdots & b_{kp} \end{bmatrix} \begin{array}{c} \uparrow \\ \\ \downarrow \end{array} m,
$$

$$
\longleftarrow p \longrightarrow \quad \longleftarrow p \longrightarrow
$$

$$
= c^p(A_{\bullet k}) * r^m(B_{k \bullet}),
$$

where, for any column vector x of dimension q, we define $c^p(x)$ to be the $q \times p$ matrix whose columns are identical to the vector x (repeated p times), while $r^m(x^T)$ is the $m \times q$ matrix whose rows are identical to the row vector x^T. Provided that (for each $k = 1, \dots, n$) these two matrices can be formed from A and B at small cost, the calculation of

$$
C = AB = [c^p(A_{\bullet 1}) * r^m(B_{1 \bullet})] + \cdots + [c^p(A_{\bullet n}) * r^m(B_{n \bullet})]
$$

is reduced to n applications of $*$.

Unlike the serial computation, where each c_{ij} is computed sequentially, here a contribution towards every c_{ij} is calculated at each stage, that is, at each application of $*$.

As pointed out above, this procedure is commonly referred to as the *outer-product algorithm*—for the obvious reason that, in the conventional serial algorithm, the inner-product loop is the innermost loop, whereas in this case the outer two loops, which embody the parallelism of the algorithm, are 'swapped' with the inner loop.

Case 2 $\min(m, n, p) = m$. The ith row $C_{i \bullet}$ of $C = [C_{1 \bullet}, \dots, C_{m \bullet}]$ has elements $c_{ij} = \sum_{k=1}^{n} a_{ik}b_{kj}$, and is therefore the vector sum of the rows of the matrix

$$\uparrow \begin{bmatrix} a_{i1} \cdots a_{i1} \\ \vdots \quad \vdots \\ a_{in} \cdots a_{in} \end{bmatrix} * \begin{bmatrix} b_{11} \cdots b_{1p} \\ \vdots \quad \vdots \\ n_{n1} \cdots b_{np} \end{bmatrix} \uparrow$$

$$\leftarrow p \longrightarrow \quad \leftarrow p \longrightarrow$$

which is $\Sigma^{\mathrm{r}} [c^p(A_{\bullet i}^{\mathsf{T}}) * B]$, where $A_{\bullet i}^{\mathsf{T}}$ is the ith column of A^{T}.

The computation is therefore carried out as

$$C = AB = \mathrm{r}\big(\Sigma^{\mathrm{r}} [c^p(A_{\bullet 1}^{\mathsf{T}}) * B], \ldots, \Sigma^{\mathrm{r}} [c^p(A_{\bullet m}^{\mathsf{T}}) * B]\big),$$

where one row of C is computed at a time. There are m applications of the operation $*$, and, for each of these, there is one vector summation of the rows of an $n \times p$ matrix, the resulting vector being assigned as a row of C.

Case 3 $\min(m, n, p) = p$. This is similar to Case 2, except that the resulting matrix C is formed by columns.

The jth column $C_{\bullet j}$ of $C = [C_{\bullet 1}, \ldots, C_{\bullet p}]$ is computed as $\Sigma^{\mathrm{c}} [A * \mathrm{r}^m(B_{j \bullet}^{\mathsf{T}})]$, where $B_{j \bullet}^{\mathsf{T}}$ is the jth row of B^{T}.

The computation of C is therefore carried out as

$$C = AB = \mathrm{c}\big(\Sigma^{\mathrm{c}} [A * \mathrm{r}^m(B_{1 \bullet}^{\mathsf{T}})], \ldots, \Sigma^{\mathrm{c}} [A * \mathrm{r}^m(B_{p \bullet}^{\mathsf{T}})]\big).$$

There are p applications of the operation $*$, and, for each of these, there is one vector summation of the columns of an $m \times n$ matrix, the resulting vector being assigned as a column of C.

(c) $N < \max(mn, np, mp)$

In this case, the matrices need to be partitioned into smaller units, where each unit is dealt with in parallel; for example, if the matrices $A, B, \ldots, H$ are all of order $n \times n$, and with n^2 processors available, the most obvious method of multiplying matrices of order $(2n)^2$ is as follows:

$$\begin{bmatrix} A & B \\ C & D \end{bmatrix} \begin{bmatrix} E & F \\ G & H \end{bmatrix} = \begin{bmatrix} AE + BG & AF + BH \\ CE + DG & CF + DH \end{bmatrix} \qquad (3.2)$$

where each product in the right-hand side of eqn (3.2) is computed in parallel. There are two general techniques for tackling such problems, which are discussed in Sections 2.9 and 2.10.

Finally, if $N \geqslant n$, then some parallelism can be obtained by performing the inner products using recursive doubling, thereby reducing the applications of termwise multiplications to n^2, with $\log_2 n$ termwise additions for each multiplication.

In general, the choice of any one of the aforementioned methods depends strongly on the ratio of computation to communication. In the case of fine granularity (as found on the DAP), where each instruction constitutes a logical operation, the above characterization, with each processor capable of performing arithmetic on its local data, seems most appropriate. However, for coarse granularity (as found on the T-rack with transputers), each processor may be assigned a set of calculations rather than individual operations, with fewer messages passing to other processors, and this may well be suitable for some implementations.

The preceding discussion is aimed at minimizing the number of termwise multiplications. The cost of addition may also be significant, particularly the cost of the $\lceil \log_2 n \rceil$-step recursive-doubling algorithm for adding n numbers—see Section 2.5 (recall that $\lceil x \rceil$ denotes the smallest integer greater than or equal to x).

The above analysis does not take account of the cost of accessing operands. These costs frequently represent a significant restriction on the performance of an algorithm, even if there are enough processors to match the dimension of the problem. In practice, operands may be accessed via a shared memory, resulting in contention for memory cycles, or (as on the DAP) each processor may have a private memory so that data movement between the processors must be taken into account. For the latter case, Gentleman (1978) has conducted research into the data movement required for matrix multiplication, and for the inversion of a matrix on a lattice of interconnected processors. His analysis confirms that data movement—and not arithmetic operations—is often the limiting factor in the performance of algorithms, and corroborates the author's view (see Chapter 1, Section 1.10) that conventional complexity analyses for parallel computations commonly ignore the details of machine structure, which can often result in misleading conclusions. In brief, it suggests more attention should be paid to the hardware characteristics of a particular implementation. A recent paper by Parkinson (1987) discusses these issues in detail, with a number of interesting examples.

Parallel matrix multiplication is a topic so vast that the above discussion constitutes no more than a 'drop in the ocean'. To do justice to it would require a comprehensive analysis with reference to the latest parallel machines. Perhaps a young enthusiast will consider writing a book on this subject alone.

2.4 Parallel evaluation of arithmetic expressions

Arithmetic expressions are central to any type of computation, and it is pertinent therefore to consider their evaluation on a parallel computer.

An arithmetic expression is built up from variables $x_1, \ldots, x_n$ by means of the operations of addition, subtraction, multiplication, and division. Two expressions are *equivalent* if they take the same value for every assignment of values to the variables.

On a parallel computer, the evaluation of an arithmetic expression E is based on the selection of an equivalent expression $\bar{E}$ for which several operations can be carried out simultaneously.

As a simple example, we consider

$$E = (x_1 x_2 + x_3)x_4 + x_5. \tag{4.1}$$

The order of evaluation of (4.1) on a serial computer may be indicated by

$$E = ((((x_1 \cdot x_2) + x_3) \cdot x_4) + x_5).$$

The equivalent

$$\bar{E} = x_1 x_2 x_4 + x_3 x_4 + x_5,$$

suitable for parallel computation, is evaluated as

$$\bar{E} = (((x_1 \cdot x_2) \cdot x_4) + ((x_3 \cdot x_4) + x_5)),$$

where the rule is that we compute *all* the inner brackets at step 1, do all 'next' brackets at step 2, etc. The serial and parallel evaluations using E and $\bar{E}$ respectively are shown in Fig. 4.1.

In serial computation, the operations are carried out sequentially, leading to expression E as a result, as in Fig. 4.1(i). In the parallel case, at step $t = 1$, the products $x_1 x_2$ and $x_3 x_4$ are

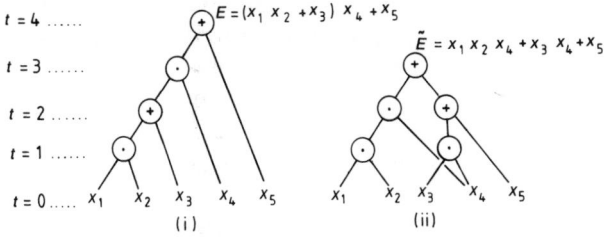

Fig. 4.1 (i) Serial and (ii) parallel evaluations of (3.1)

computed simultaneously; at step $t = 2$, the product $x_1 x_2 \cdot x_4$ and the sum $x_3 x_4 + x_5$ are computed simultaneously; finally, at step $t = 3$, the sum is computed to produce $\tilde{E}$.

We can compare the serial and parallel evaluations as follows. Let

t = number of parallel or serial steps,

p = number of processors used,

s = total number of operations performed by the algorithm.

In Table 4.1 the values of (t, p, s) are displayed for the two cases. Although the number of parallel steps is one less than the number of serial steps, the total number of individual operations is increased by one in the parallel computation. This example thus underlines the connection between optimizing the use of processors, which may otherwise lie idle, and minimizing the total number of steps.

We are clearly assuming that different operations can be carried out simultaneously, and hence in general an MIMD type of parallel machine is most appropriate. Using an SIMD

Table 4.1 Comparison of serial and parallel evaluations

	Serial evaluation, E	Parallel evaluation, $\tilde{E}$
t:	4	3
p:	1	2
s:	4	5

machine, on which additions and multiplications cannot be done simultaneously, four steps would be needed. On the other hand, for

$$E_7 = (((x_1x_2) + x_3)x_4 + x_5)x_6 + x_7,$$

with the equivalent expression

$$\tilde{E}_7 = ((((x_1x_2)(x_4x_6)) + (x_3(x_4x_6))) + ((x_5x_6) + x_7)),$$

which involves four steps on an MIMD machine but five on an SIMD machine, there is an alternative expression

$$\tilde{\tilde{E}}_7 = ((((x_1x_2) + x_3)(x_4x_6)) + ((x_5x_6) + x_7))$$

which also involves only four steps on an SIMD machine. The type of parallel machine may therefore influence the choice of equivalent expression. Several complexity results have been established with respect to the number of parallel steps, and the number of processors, for general arithmetic expressions. Brent (1973) provides a summary of early work. For further reading see Winograd (1975), Hyafil and Kung (1977), and Brent (1974).

The stability of equivalent arithmetic expressions is the subject of continual investigation. For floating-point computation, the associative law does not hold, and the substitution of $\tilde{E}$ may therefore lead to numerical instability. Ronsch has shown that, for an arbitrary arithmetic expression E, with distinct variables, the relative accumulated rounding error R is bounded by

$$R \leqslant R_1R_2,$$

where R_1 denotes the relative error due to inaccurate input data and R_2 denotes the number of $+$ and $-$ operations in E. Assuming that the original expression E is stable, it is the transformation process $E \rightarrow \tilde{E}$ that may introduce numerical instability. For further details see Feilmeier (1982).

The evaluation of arithmetic expressions in parallel and the exploitation of concurrency in data-flow architectures are closely related. In the former case, parallelism may be specified using a 'tree' structure as indicated, for example, in Fig. 4.1(ii), for evaluating $\tilde{E}$. In the latter case, concurrency may be expressed using a data-dependence graph. We shall briefly consider exploitation of parallelism on a data flow system.

The conventional view of a program—a list of instructions that manipulate data in a fixed storage location in a defined sequence—is abandoned. Instead, the relationship between data and operations is viewed in terms of a data-dependence graph, which shows how instructions depend on data. Consider, for example, a program consisting of the following statements written in some high-level language:

$$g = a + d$$
$$h = c + d$$
$$i = e + f$$
$$k = g \cdot h \cdot i.$$

These may be written as:

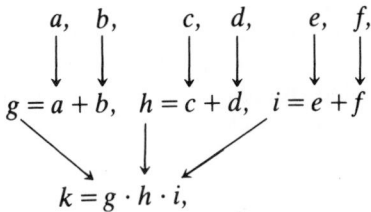

which illustrates the potential parallelism across, and the enforced sequence down, the page. The tail of a directed edge shows where the variable is assigned, and the head shows where it is consumed (i.e. appears on the right-hand side of an assignment statement). In this graphical form, the variable names become superfluous, although as an aid to understanding the sequence of computations involved they may be specified alongside the edges as shown in Fig. 4.2.

This is now referred to as a statement-level data-dependence

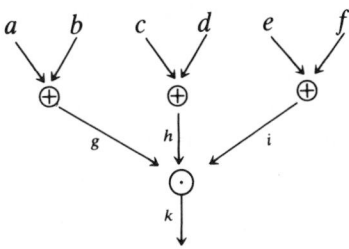

Fig. 4.2 Data-dependence graph

Fig. 4.3   Data-dependence subgraphs

graph (and is comparable to Fig. 4.1(ii) for evaluating $\bar{E}$). In order to exploit parallelism at the instruction level, further decomposition of the program may become necessary. In the above example, however, the ordering of evaluation at the multiplication stage is not fully specified. This may be resolved by one of the subgraphs in Fig. 4.3.

In this case, the compiler can choose either of the alternatives. In general, care should be taken when the order of evaluation is important. A detailed discussion on the exploitation of parallelism by the data-flow approach is provided by Gurd (1982).

2.5   Recursive doubling

If ∘ is an associative operation on pairs of mathematical objects (numbers, vectors, matrices, etc.), thus

$$(a \circ b) \circ c = a \circ (b \circ c),$$

then the 'product'

$$a_1 \circ a_2 \circ \cdots \circ a_n$$

is uniquely defined independently of the bracketing; for example,

$$((a_1 \circ a_2) \circ a_3) \circ a_4 = (a_1 \circ a_2) \circ (a_3 \circ a_4).$$

The left-hand side represents the natural way of calculating the product as in Fig. 5.1(i); the right-hand side is an alternative method, such that on a parallel machine the operations in the brackets can be carried out simultaneously, as in Fig. 5.1(ii).

Computations within each level are performed in parallel, and (in general) if the size of the set of objects is n, then the result is produced in $\lceil \log_2 n \rceil$ steps. This simple idea is the basis of recursive doubling, whereby the total computation is repeatedly divided into two separate computations of equal complexity that

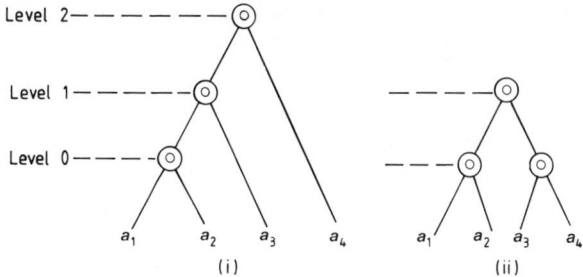

Fig. 5.1 (i) Serial evaluation, and (ii) parallel evaluation using recursive doubling

can be executed in parallel. The natural means of carrying out these operations is to use a binary tree interconnection of processors (see Chapter 1, Section 1.7). If we take $2^3 = 8$ numbers, then the calculations may be arranged as shown in Fig. 5.2, where each of the processors $P_1,..., P_7$ performs an associative operation $\circ$ on distinct pairs of objects chosen from $a_1,..., a_8$.

At step 1 P_4: $a_1 \circ a_2$, store result as x_1,
            P_5: $a_3 \circ a_4$, store result as x_2,
            P_6: $a_5 \circ a_6$, store result as x_3,
            P_7: $a_7 \circ a_8$, store result as x_4.
At step 2 P_2: $x_1 \circ x_2$, store result as x_5,
            P_3: $x_3 \circ x_4$, store result as x_6.
At step 3 P_1: $x_5 \circ x_6$, output result.

Computations within each step are performed simultaneously, and in three steps the result appears in P_1.

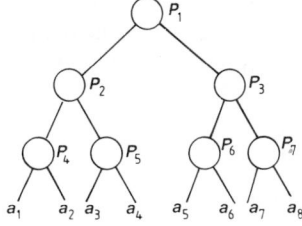

Fig. 5.2 Recursive doubling on a binary tree of the fourth order

The technique of recursive doubling is very useful in practice. We cited the example of the addition of a set of numbers in Chapter 1, Section 1.7, where the associative operator is $+$. Similarly, the product of p numbers $n_1, \ldots, n_p$ can be computed in parallel by taking $\circ$ to be $*$. For finding the maximum or minimum of a set of numbers, the operation $\circ$ is replaced by a comparator $\vee$ or $\wedge$. The objects chosen thus far have been scalars; in the following examples, the operations are performed on vectors and matrices.

Evaluation of recurrence relations

In a number of cases, the solution of a problem reduces to the evaluation of the last term, or all terms, of a recurrently defined sequence, and this turns out to be ideally suited to recursive doubling.

Suppose we are given the entries $\alpha_1, \ldots, \alpha_n$ and $\beta_2, \ldots, \beta_n$ of the tridiagonal matrix

$$
B = \begin{bmatrix}
\alpha_1 & \beta_2 & & & \\
\beta_2 & \alpha_2 & \beta_3 & & \\
& \beta_3 & \alpha_3 & \ddots & \\
& & \ddots & \ddots & \beta_n \\
& & & \beta_n & \alpha_n
\end{bmatrix}
$$

and we wish to compute the determinant of B. The determinant f_r of the rth leading minor satisfies

$$f_1 = \alpha_1, \qquad f_2 = \alpha_1 \alpha_2 - \beta_2^2, \qquad f_r = \alpha_r f_{r-1} - \beta_r^2 f_{r-2} \quad (r = 3, \ldots, n).$$

This is a special case of the two-term linear recurrence

$$x_i = a_i x_{i-1} + b_i x_{i-2} \quad (i = 2, \ldots, n), \tag{5.1}$$

to be solved for $x_2, \ldots, x_n$, given the values of x_0 and x_1, and the a_i and b_i. At first sight, the parallelism in (5.1) is not obvious. Let us reformulate the equation in matrix–vector notation. Write

$$y_1 = \begin{bmatrix} x_1 \\ x_0 \end{bmatrix}, \qquad y_i = \begin{bmatrix} x_i \\ x_{i-1} \end{bmatrix} \quad (i = 2, \ldots, n).$$

Then (5.1) can be written

$$y_i = M_i y_{i-1}, \qquad M_i = \begin{bmatrix} a_i & b_i \\ 1 & 0 \end{bmatrix} \quad (i = 2, \ldots, n);$$

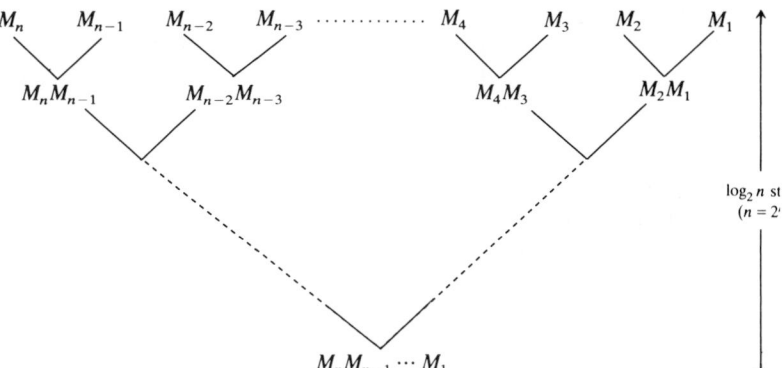

Fig. 5.3 Evaluation of a second-order linear recurrence formula

therefore

$$\begin{bmatrix} x_n \\ x_{n-1} \end{bmatrix} = y_n = M_n M_{n-1} \cdots M_2 \begin{bmatrix} x_1 \\ x_0 \end{bmatrix}, \tag{5.2}$$

and it is sufficient to calculate the product of 2×2 matrices in (5.2). This may be performed as shown in Fig. 5.3, where for simplicity it is assumed that $n = 2^k$.

The computation can be completed in $\log_2 n$ steps, provided that the 2×2 matrix multiplications can be carried out in parallel. Each matrix multiplication can be accomplished with two termwise multiplications by the method of Section 2.3, and thus a general two-term linear recurrence takes a time proportional to $2 \log_2 n$ termwise multiplications, provided that the number of processors is sufficiently large. Similarly, a three-term linear recurrence

$$x_i = c_i x_{i-1} + b_i x_{i-2} + a_i x_{i-3} \quad (i = 3, \ldots, n)$$

can be evaluated in a time proportional to $3 \log_2 n$ multiplications, and this result clearly generalizes for p-term recurrences. As well as the multiplications, various additions must, of course, be made, and partial sums assigned to the correct positions. On some implementations, these costs may represent a significant limitation.

A two-term recurrence of the form

$$x_i = a_i x_{i-1} + b_i x_{i-2} + c_i$$

can be expressed as

$$
\begin{bmatrix} x_i \\ x_{i-1} \\ 1 \end{bmatrix} = \begin{bmatrix} a_i & b_i & c_i \\ 1 & 0 & 0 \\ 0 & 0 & 1 \end{bmatrix} \begin{bmatrix} x_{i-1} \\ x_{i-2} \\ 1 \end{bmatrix}
$$

and treated as a three-term recurrence.

Certain non-linear recurrences can be reduced to linear ones, which can then be solved by the above technique, with the help of suitable changes of variable. For example, the first-order non-linear recurrence

$$x_0 \quad \text{given}, \qquad x_i = a_i + b_i/x_{i-1}, \tag{5.3}$$

if both sides are multiplied by $x_0 x_1 \cdots x_{i-1}$, becomes

$$y_0 \quad \text{given}, \qquad y_i = a_i y_{i-1} + b_i y_{i-2} \tag{5.4}$$

(a second-order linear recurrence), with $y_i = x_0 x_1 \cdots x_i$. The more general linear–fractional case

$$x_0 \quad \text{given}, \qquad x_i = (a_i x_{i-1} + b_i)/(c_i x_{i-1} + d_i) \tag{5.5}$$

reduces to the form (5.3) if we put $X_i = c_{i+1} x_i + d_{i+1}$.

These recurrences arise in a number of applications, including the evaluation of continued fractions and spline functions and the solution of tridiagonal systems and certain differential equations.

Another, closely related, method can be used to solve (5.5) using recursive doubling. If we substitute the expression $x_1 = (a_1 x_0 + b_1)/(c_1 x_0 + d_1)$ in the corresponding formula for x_2 in terms of x_1, we obtain the following formula for x_2 as a function of x_0:

$$x_2 = \frac{(a_2 a_1 + b_2 c_1)x_0 + a_2 b_1 + b_2 d_1}{(c_2 a_1 + d_2 c_1)x_0 + c_2 b_1 + d_2 d_1}, \tag{5.6}$$

and this process can be continued. Suppose that the final result is

$$x_n = (\alpha x_0 + \beta)/(\gamma x_0 + \delta). \tag{5.7}$$

Then it is clear from (5.6) that the matrix of coefficients $\begin{bmatrix} \alpha & \beta \\ \gamma & \delta \end{bmatrix}$ is simply the *matrix product*

$$
\begin{bmatrix} \alpha & \beta \\ \gamma & \delta \end{bmatrix} = \begin{bmatrix} a_n & b_n \\ c_n & d_n \end{bmatrix} \begin{bmatrix} a_{n-1} & b_{n-1} \\ c_{n-1} & d_{n-1} \end{bmatrix} \cdots \begin{bmatrix} a_1 & b_1 \\ c_1 & d_1 \end{bmatrix}.
$$

This product can be calculated using the recursive doubling technique, and x_n can be found from (5.7).

The method applies to any recurrence problem of the form

$$x_0 \quad \text{given}, \qquad x_i = f_i(x_{i-1}), \tag{5.8}$$

where the functions f_i are given members of a class $\mathscr{F}$ of functions, each depending on certain numerical coefficients $a_i, b_i, \ldots$, provided that the composite $x \mapsto f(g(x))$ of two functions of the class $\mathscr{F}$ is again a function of the same class, with conveniently calculable coefficients. The solution of (5.8) is evidently

$$x_n = f_n(f_{n-1}(\cdots f_1(x_0) \cdots)), \tag{5.9}$$

and the procedure is to calculate (the coefficients of) the composite function

$$f_n \circ f_{n-1} \circ \cdots \circ f_1 = x \mapsto f_n(f_{n-1}(\cdots f_1(x) \cdots)) \tag{5.10}$$

and then evaluate it at $x = x_0$. For the calculation of the function (5.10), recursive doubling is clearly appropriate; for instance, if $n = 8$ we can compute it in three parallel steps as follows:

$$((f_8 \circ f_7) \circ (f_6 \circ f_5)) \circ ((f_4 \circ f_3) \circ (f_2 \circ f_1))$$

The application of recursive doubling to recurrence problems of this type is due to Kogge. For more detail, see Kogge (1973) and Feilmeier (1982), and for further discussion of some of the above topics, see also Stone (1973). Parallel evaluation of a system of linear recurrences is described in Section 4.3 of Chapter 4.

2.6 Cyclic odd–even reduction

The technique of cyclic odd–even reduction, dating back to the 1960s (Hockney 1965), can be used in a wide variety of computations which reduce to the solution of recurrence relations. It appears to be applicable to the same class of problem as recursive doubling (Section 2.5), and facilitates a similar increase in computation speed. The idea of reformulating computations to enhance the inherent parallelism of the sequential algorithm is of particular importance here.

The technique can be illustrated with the solution of a first-order recurrence

$$x_i = a_i + b_i x_{i-1} \tag{6.1}$$

where a_i and b_i are known ($i = 1, \ldots, n$) and $x_1, \ldots, x_n$ are to be calculated, given $x_0 = \alpha$. We set $a_0 = \alpha$, and $a_i = 0$ for $i < 0$, with $b_i = 0$ for $i \leqslant 0$. Then (6.1) will also hold for $i \leqslant 0$, with $x_i = 0$ for $i < 0$.

From (6.1), and (6.1) with i replaced by $i - 1$, we have

$$x_i = a_i + b_i(a_{i-1} + b_{i-1}x_{i-2}) = a_i^{(1)} + b_i^{(1)}x_{i-2} \quad \text{(say)}; \tag{6.2}$$

from (6.2), and (6.2) with i replaced by $i - 2$, we have

$$x_i = a_i^{(1)} + b_i^{(1)}(a_{i-2}^{(1)} + b_{i-2}^{(1)}x_{i-4}) = a_i^{(2)} + b_i^{(2)}x_{i-4}. \tag{6.3}$$

Continuing in this way, x_i can be expressed successively in terms of x_{i-1}, x_{i-2}, x_{i-4}, $x_{i-8}, \ldots$; after at most k steps, where $k = \lceil \log_2 n \rceil$, the index $i - 2^k$ will be negative or zero, and therefore $b_{i-2}^{(k)} = 0$ and x_i is equal to $a_{i-2^k}^{(k)}$. The algorithm is thus as follows.

With $a_i^{(0)} = a_i$ and $b_i^{(0)} = b_i$ for all i:

$$a_i^{(p)} = a_i^{(p-1)} + b_i^{(p-1)}a_{i-\theta(p)}^{(p-1)}, \qquad b_i^{(p)} = b_i^{(p-1)}b_{i-\theta(p)}^{(p-1)},$$

where $\theta(p) = 2^{p-1}$. The result appears as

$$x_i = a_i^{(k)}, \qquad k = \lceil \log_2 n \rceil.$$

This technique can also be applied to the solution of a diagonally dominant tridiagonal system of n linear equations in n unknowns:

$$Ax = y, \tag{6.4}$$

where A and y are given, and x is to be found. In matrix-vector notation, (6.4) can be written as

$$
\begin{bmatrix}
b_1 & c_1 & & & & & 0 \\
a_2 & b_2 & c_2 & & & & \\
& a_3 & b_3 & c_3 & & & \\
& & & \ddots & \ddots & \ddots & \\
0 & & & & a_{n-1} & b_{n-1} & c_{n-1} \\
& & & & & a_n & b_n
\end{bmatrix}
\begin{bmatrix}
x_1 \\ x_2 \\ x_3 \\ \vdots \\ x_{n-1} \\ x_n
\end{bmatrix}
=
\begin{bmatrix}
y_1 \\ y_2 \\ y_3 \\ \vdots \\ y_{n-1} \\ y_n
\end{bmatrix}.
$$

Consider the three equations involving x_i:

$$a_{i-1}x_{i-2} + b_{i-1}x_{i-1} + c_{i-1}x_i = y_{i-1}, \tag{6.5}$$

$$a_i x_{i-1} + b_i x_i + c_i x_{i+1} = y_i, \tag{6.6}$$

$$a_{i+1}x_i + b_{i+1}x_{i+1} + c_{i+1}x_{i+2} = y_{i+1}. \tag{6.7}$$

By subtracting suitable multiples of eqns (6.5) and (6.7) from (6.6), we can obtain an equation of the form

$$a_i^{(1)}x_{i-2} + b_i^{(1)}x_i + c_i^{(1)}x_{i+2} = y_i^{(1)}. \tag{6.8}$$

By similar manipulation of three equations of the form (6.8) we obtain

$$a_i^{(2)}x_{i-4} + b_i^{(2)}x_i + c_i^{(2)}x_{i+4} = y_i^{(2)},$$

and so on. Setting $x_i = a_i = c_i = y_i = 0$ and $b_i = 1$ when i is outside the relevant range, then, after at most k steps, where $k = \lceil \log_2 n \rceil$, we reach an equation

$$b_i^{(k)}x_i = y_i^{(k)}, \tag{6.9}$$

giving the value of x_i.

Diagonal dominance means here that

$$b_i > |a_i| + |c_i|$$

for all i, and it is easy to see that this property is inherited by the later coefficients:

$$b_i^{(k)} > |a_i^{(p)}| + |c_i^{(p)}| \quad (p = 1, \ldots, k).$$

In particular, $b_i^{(k)}$ in eqn (6.9) is positive.

Odd–even reduction for banded linear equations and for pentadiagonal matrices is discussed by Rodrigue *et al.* (1979) and Madsen and Rodrigue (1977) respectively.

2.7 Bit-level algorithms

Some older readers may remember serial machines like the Pegasus, or the Elliot 903, which reduced the cost of the arithmetic unit by streaming the bits one after another through a single adder. Handling the bits of a word in parallel was one of the earliest forms of parallelism to be used successfully in general-purpose computing.

In a bit-organized system, each processor can operate on bits of an operand during each memory cycle. In contrast, in a word-organized system, each processor operates on a word—typically 32 or 64 bits (see Chapter 1, Section 1.3). Bit-oriented algorithms are particularly suited to high interprocessor bandwidth and zero latency time. In this section, we examine some algorithms that can be formulated at bit level and can also exploit the parallel nature of the machine. The memory can be thought of as consisting of several bit planes, with the data (numbers) being stored vertically downwards, as shown in Fig. 7.1. The operations on the entire bit plane can be done in parallel; thus, if we have n p-bit numbers represented vertically, then we can interrogate the rth bit $(1 \leqslant r \leqslant p)$ of all numbers simultaneously.

One important consequence of the vertical representation of data is that algorithms operating on bit planes are independent of the number of data items, provided that enough processors are available. The time taken depends only on the bit-length of the numbers. The concept of vertical mapping has greatly influenced the design of some parallel algorithms and is illustrated below by means of two examples: that of finding the maximum of a set of non-negative integers, and that of adding a set of numbers.

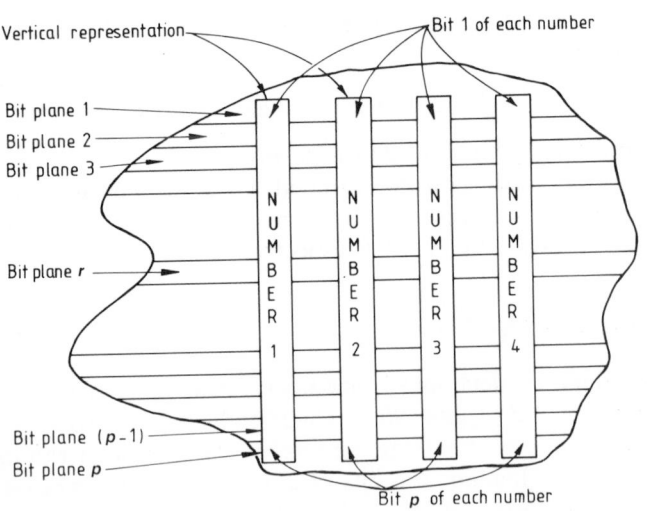

Fig. 7.1 Vertical representation of four p-bit numbers

The procedure for finding the maximum element(s) of a set of non-negative integers, represented as binary numbers of the same length stored vertically downwards, is as follows.

Begin by regarding all the numbers as candidates. Then examine the planes successively, at each stage by means of a logical AND operation removing from the current candidate set all numbers with a zero in the plane under examination, unless this would eliminate all candidates, in which case the candidate set is left unchanged. At the end of the process, precisely the maximum elements (or element) remain as candidates.

In Table 7.1 we illustrate the procedure, as applied to the numbers 6, 8, 12, 13, 7, and 13; asterisks indicate the numbers remaining in the candidate set at each stage.

We next consider the process of summation of N numbers:

$$y = \sum_{i=1}^{N} x_i \quad (i = 1, \dots, N).$$

By the recursive-doubling technique (Section 2.5), it was seen that, on an N-processor system, the time taken to sum N numbers is proportional to $\log_2 N$, provided that there is no overhead due to data movement. During the computation, processor utilization diminishes: 50 per cent at the first step, 25 per cent at the second step, $12\frac{1}{2}$ per cent at the third, and so on.

Table 7.1 Parallel manipulation on bit planes to find maximum elements

6	8	12	13	7	13
*	*	*	*	*	*
0	1	1	1	0	1
	*	*	*		*
1	0	1	1	1	1
		*	*		*
1	0	0	0	1	0
		*	*		*
0	0	0	1	1	1
			*		*

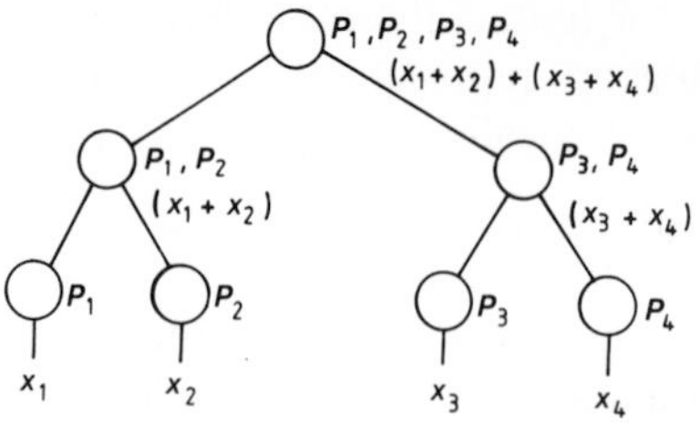

Fig. 7.2 Sharing of processors to perform addition

In a bit-organized system the arithmetic operations are shared among the processors, so that maximum possible utilization of processors is achieved; for example, at the kth step of the recursive-doubling process, one of the addition operations $x_i^{(k)} + x_j^{(k)}$ is shared among the available processors: one group of processors may operate on the most significant bits, whereas the other group may operate on the least significant bits.

Let us consider the addition $x_1 + x_2 + x_3 + x_4$ with four processors P_1, P_2, P_3, P_4, displayed in Fig. 7.2. Assume that initially x_1, x_2, x_3, x_4 are stored in P_1, P_2, P_3, P_4, respectively.

At Step 1: P_1 and P_2 both contribute to the addition $y_1 = x_1 + x_2$,
$\quad\quad\quad\quad$ P_3 and P_4 both contribute to the addition $y_2 = x_3 + x_4$.
At Step 2: P_1, P_2, P_3, and P_4 contribute to the addition $y_1 + y_2$.

In this way all processors are fully utilized; indeed, on the DAP, a number of algorithms have been constructed using this principle, whereby several processors perform arithmetic on bits of data items rather than individual processors performing arithmetic on individual data items (as in word-based systems).

The timing characteristics of bit-organized systems have led to some interesting conclusions. On a serial computer, the computational time does not depend on the bit complexity of the operation, because the operations are performed on fixed-size words. The dependence of operation time on the number of bits in the operand is thus a concept unfamiliar to most users. The consequences of this feature are evident if we examine the performance of some common operations on the DAP, as shown in Table 7.2. (In fact on a recently developed Mini-DAP, with $32 \times 32 = 1024$ processors, a speed-up factor of between 1 and 2

Table 7.2 Timing, in μsec, of some DAP operations (bit-organized system)

Operation	Time (μsec)	Number of results
16-bit data movement ($I = J$)	20	4096
single-bit Boolean (L AND M)	2	4096
sign change of 32-bit number	1	4096
16-bit fixed-point addition ($I = J + K$)	20	4096
32-bit fixed-point addition ($A = B + C$)	38	4096
32-bit floating-point addition/subtraction ($X = Y \pm Z$)	173	4096
32-bit floating-point multiplication ($X = Y * Z$)	270	4096
32-bit floating-point division ($X = Y/Z$)	384	4096
32-bit square of a number ($X = Y ** 2$)	160	4096
32-bit square root ($X = $ SQRT(Y))	190	4096
32-bit maximum value from a set	56	1
32-bit sum of a set of numbers	450	1
32-bit floating-point evaluation of SIN/COS function	860	4096
32-bit floating-point logarithm (to base e)	322	4096
32-bit floating-point exponential	413	4096

is obtained over the DAP figures for floating-point operations; for example for +, *, /, **2, and SQRT, the timings are 135, 210, 291, 107, and 136 (μsec) respectively for 32-bit arithmetic.) Table 7.2 shows that $X_i = $ SQRT(Y_i) $(i = 1,..., 4096)$ is cheaper to evaluate than $X_i = Y_i * Z_i$ $(i = 1,..., 4096)$, which is unlikely to be true on most currently available serial and vector machines. Hence, contrary to common belief, it is not disastrous to utilize square roots, or indeed the maximal function.

A brief discussion of bit-organized architecture is included in Chapter 1, Section 1.3; a more detailed description can be found in Parkinson (1982b). To reiterate the important points: fixed-point is much faster than floating-point arithmetic; short-precision is faster than long-precision arithmetic; logical operations are very fast compared to arithmetic; finally, the *relative* timings of the operations are very different from those measured on serial machines.

2.8 Bit-level algorithms for numerical functions

In this section, we discuss the calculation of standard numerical functions such as e^x, $\log_e x$, $\sin x$, and $\cos x$. As the heading suggests, the general approach is to develop algorithms that process bits of all the operands simultaneously. The schemes essentially involve multiplication or division in binary, and therefore have the advantage of taking a time comparable to that of a single multiplication or division instruction. Also, as indicated in Chapter 1 (Section 1.3), they can be adjusted to a required degree of accuracy. It is remarkable that algorithms for some of the most powerful and sophisticated computers have features similar to those in ordinary pocket calculators.

In simplified form, the method can be illustrated by the calculation of e^x in parallel for a set of values x in the range $0 < x < 1$. The values of e^{δ_n} for $\delta_n = 2^{-n}$ $(n = 1,..., N)$ are stored in a look-up table.

If $x = \cdot\, b_1 b_2 \cdots b_N$ in binary (to the required accuracy), then

$$x = b_1 \delta_1 + b_2 \delta_2 + \cdots + b_N \delta_N, \qquad e^x = e^{b_1 \delta_1} e^{b_2 \delta_2} \cdots e^{b_N \delta_N}.$$

Hence if we compute E_n $(n = 1, 2, ...)$ by

$$E_0 = 1, \qquad E_n = \begin{cases} E_{n-1} e^{\delta_n} & \text{if } b_n = 1, \\ E_n & \text{if } b_n = 0, \end{cases}$$

then $E_N = e^x$. This procedure is particularly suited to a bit-oriented SIMD machine, the values of x being stored vertically (Fig. 7.1).

In general, for a relation $Y = F(X)$, we wish to calculate, in parallel, a set of values of Y corresponding to given X's (the direct problem), or values of $X = F^{-1}(Y)$ corresponding to given Y's (the inverse problem). The basis of the method can be described as follows.

Step 1   Scale X to x, and/or Y to y, in order to bring the ranges of coordinate values into convenient intervals, in which y will usually vary monotonically, smoothly, and neither too rapidly nor too slowly with respect to x. Call the resulting curve $y = f(x)$.

Step 2   Derive an exact formula for $f(x + \Delta x)$ as a function of x, $f(x)$, and Δx.

Step 3   Select a starting point (ξ_0, η_0) on the curve, and a fixed set of increments $\delta_1,..., \delta_N$ such that all the given x's can be reached (to the required accuracy) from ξ_0 by using these increments:

$$x = \xi_0 + \varepsilon_1\delta_1 + \varepsilon_2\delta_2 + \cdots + \varepsilon_N\delta_N.$$

Here the ε_i will be chosen from a specified two-element set T, usually $\{0, 1\}$ or $\{-1, 1\}$. It must also be possible to calculate $f(x + \Delta x)$ by the formula of Step 2, when $\Delta x = \varepsilon_n\delta_n$, if necessary by means of a look-up table of limited size.

In a direct problem, the algorithm is as follows: we define

$$x_0 = \xi_0, \quad y_0 = \eta_0, \quad x_n = x_{n-1} + \varepsilon_n\delta_n,$$

where $\varepsilon_n \in T$ is so chosen that x_n is a better approximation to x than x_{n-1}, and let

$$y_n = f(x_n).$$

Then $y_N = f(x_N) = f(x)$ to the required accuracy.

In an inverse problem, we treat the numbers $x_{n-1} + \varepsilon_n\delta_n$ ($\varepsilon_n \in T$) as trial values, and select ε_n so that $f(x_n) = f(x_{n-1} + \varepsilon_n\delta_n)$ is a better approximation to y than $f(x_{n-1})$. Then $y_N = y$ and $x_N = f^{-1}(y)$ to the required accuracy.

Choice of increments

This is governed not only by the need to facilitate the determination of ε_n and the computation of y_n at each stage, but also the requirement that all x's in the chosen range should be accessible. When $T = \{0, 1\}$, the following well-known theorem gives a necessary and sufficient condition for the latter to be true in the limiting case $N \to \infty$, and provides a useful guide in the finite case.

Let $\delta_1 \geqslant \delta_2 \geqslant \cdots \to 0$. Then every x satisfying $0 \leqslant x \leqslant \sum_{n=1}^{\infty} \delta_n$ can be expressed in the form $\sum_{n=1}^{\infty} \varepsilon_n\delta_n$, with $\varepsilon_n \in \{0, 1\}$, if and only if $\sum_{n=N}^{\infty} \delta_n \geqslant \delta_{N-1}$ for all $N = 2, 3, \ldots$.

We next discuss the calculation of square roots, exponentials and logarithms, and sines and cosines (for further details see Gostick 1979).

Square root

This is a particularly simple example of an inverse problem. To find the square root of y, where $0 < y < 1$, take $x_0 = 0$, $T =$

$\{0, 1\}$, $\delta_n = 2^{-n}$ $(n = 1,..., N)$, and $r_0 = y$. At the nth stage we have an approximation x_n to $y^{\frac{1}{2}}$, and an error $r_n = y - x_n^2$. Then $x_n = x_{n-1} + \varepsilon_n \delta_n$, and so if $\varepsilon_n = 1$, then

$$e_n = y - x_n^2 = r_{n-1} - (2x_{n-1} + \delta_n)\delta_n. \tag{8.1}$$

Thus we calculate the value of (8.1); if this gives $r_n > 0$, then we take $\varepsilon_n = 1$ and $x_n = x_{n-1} + \delta_n$; otherwise $x_n = x_{n-1}$ and $r_n = r_{n-1}$. The final x_N is the value of $y^{\frac{1}{2}}$ to the required accuracy.

Arithmetic is not required in the calculation of the error term $(2x_{n-1} + \delta_n)\delta_n$ in (8.1): the product $2x_{n-1}$ in binary is simply x_{n-1} shifted one place to the left, producing a 0 at the right-hand end; addition of δ_n replaces 0 by 1 at the right-hand end; and finally, multiplication by δ_n entails a shift of n places to the right. The number of operations required is about half that of a division operation to the same precision.

Exponential and logarithm

The method for computing e^x described above can clearly be subsumed under the general scheme, the exact formula of Step 2 being in this case $f(x + \Delta x) = f(x)f(\Delta x)$.

One variant used in practice, still with $\varepsilon_n \in \{0, 1\}$, is to take $\xi_0 = 0$, $\eta_0 = 1$ and $\delta_n = \log_e(1 + 2^{-n})$, these values being stored, and the range for x being say $[0, \frac{1}{2}]$.

The formula for y_n when $\varepsilon_n = 1$ is then

$$y_n = y_{n-1}(1 + 2^{-n}),$$

which is easy to compute since it only requires a shift and an add at each step.

The method used for calculating the logarithms of numbers x between $\frac{1}{2}$ and 1 involves successive multiplication by fixed factors until the product reaches 1 to the required accuracy.

Let $a_n = 1 + 2^{-n}$, store the numbers $\log_e a_n$ $(n = 1,..., N)$, and define

$$x_0 = x, \qquad x_n = \begin{cases} x_{n-1}a_n & \text{if this is less than 1,} \\ x_{n-1} & \text{otherwise;} \end{cases}$$

$$\varepsilon_n = \begin{cases} 1 & \text{if } x_{n-1}a_n < 1, \\ 0 & \text{otherwise.} \end{cases}$$

Then $x_N = x \prod_{n=1}^{N} a_n^{\varepsilon_n} = 1$, and

$$y_N = \sum_{n=1}^{N} \varepsilon_n \log_e a_n = -\log_e x.$$

Sine and cosine

In this case, we make use of the identities

$$\sin(x \pm y) \equiv \cos y \, (\sin x \pm \cos x \tan y),$$
$$\cos(x \pm y) \equiv \cos y \, (\cos x \mp \sin x \tan y),$$

and of the precomputed values $\delta_n = \bar{\delta}_n$ defined by

$$\tan \bar{\delta}_n = 2^{-n}, \quad 0 < \bar{\delta}_n < \tfrac{1}{2}\pi \quad (n = 0, \dots, N).$$

This time the set T is $\{-1, 1\}$. To find the sine or cosine of a given number p (in practice we calculate sine and cosine simultaneously for a set of arguments), we proceed as follows.

First normalize p to lie in the range $0 < p < \tfrac{1}{2}\pi$; then, starting with $\sin 0 = 0$ and $\cos 0 = 1$, add or subtract successive increments of $\bar{\delta}_n$ until we have built up p. Hence, at the $(n+1)$th stage, having used $\bar{\delta}_0, \dots, \bar{\delta}_n$, the approximation to p is

$$p_{n+1} = \begin{cases} p_n + \bar{\delta}_{n+1} & \text{if } p_n < p, \\ p_n - \bar{\delta}_{n+1} & \text{otherwise.} \end{cases}$$

The corresponding approximations to $\sin p$ and $\cos p$ are as follows:

$$\begin{bmatrix} \sin p_{n+1} \\ \cos p_{n+1} \end{bmatrix} = \cos \bar{\delta}_{n+1} \begin{bmatrix} 1 & \pm 2^{-n-1} \\ \mp 2^{-n-1} & 1 \end{bmatrix} \begin{bmatrix} \sin p_n \\ \cos p_n \end{bmatrix}.$$

The terms $\cos \bar{\delta}_i$ enter into the final result only as the product

$$\cos \bar{\delta}_0 \cos \bar{\delta}_1 \cdots \cos \bar{\delta}_N,$$

which can clearly be precomputed and looked up at the end of the process. Other parts of the calculation require only shifting and adding in binary.

It is interesting to compare the execution time on the DAP for some of the above calculations, as a multiple of the time it takes to multiply two numbers as derived from Table 7.2:

square root	0.70 multiplication time
logarithm (to base e)	1.15 multiplication time
exponential	1.53 multiplication time
sine or cosine	3.10 multiplication time
add or subtract	0.64 multiplication time

(all figures refer to 32-bit floating-point arithmetic, with 24-bit mantissae).

The relative times using bit-arithmetic in vertical mode are very different from those of serial machines, which emphasizes once again that judgements of efficiency may be overturned in a parallel environment.

Squaring compared with multiplication

It was seen in Table 7.2 that squaring a number $(x = y^2)$ is faster than multiplication $(x = y * y)$. The reason for this is simple. Consider y scaled so that

$$y = \sum_{i=1}^{k} b_i 2^{-i},$$

where b_i are the binary bits. To form y^2, the b_i are multiplied as follows:

$$b_1 * (b_1, \ldots, b_k)$$
$$b_2 * (b_1, \ldots, b_k)$$
$$\vdots$$
$$b_k * (b_1, \ldots, b_k).$$

Apart from entries of type $b_i b_i$, each entry occurs twice; for example, the term in 2^{-4} has coefficient $b_1 b_3 + b_2 b_2 + b_3 b_1$, where two entries are repeated. The multiplication can, of course, be arranged so as to avoid this duplication. Squaring a number is therefore almost twice as fast as multiplication, using vertical-mode representation.

Further discussion on numeric-function evaluation is provided by Chen (1975) and Egbert (1978). Bit-oriented algorithms can also be applied to sorting and image processing (Arnot *et al.* 1982). On an array processor, vertical mapping has proved very effective in the famous Batcher sort. The basis of the method is a series of comparisons that will merge two sorted sequences of length n to give a sorted result of length $2n$ (see Chapter 3). Turner (1987) has considered the use of bit-oriented algorithms in the implementation of level-index arithmetic.

2.9 Slicing and crinkling

Slicing and crinkling are two general and rather simple techniques to accommodate a fairly large problem, using a smaller number of processors. In particular, assume a problem of size n (which may, for example, be the dimension of the matrix) that ideally requires $\alpha(n)$ processors, but for which there are only $\eta < \alpha(n)$ processors available.

We first define the term *granularity*, which is used to classify parallel computers in terms of complexity and number of processors. Machines with a large number of elementary processors, each holding a small volume of data, are *fine-grained*; those with a small number of more complex processors, each holding a large volume of data, are *coarse-grained*. In general, a coarse-grain machine has up to about 16 processors and a medium-grain up to 64; a fine-grain machine typically has at least 1000 processors. The techniques discussed below are best suited to fine granularity, with each individual processor capable only of performing arithmetic on its local data.

In the case of slicing, as the name suggests, we simply partition the problem into smaller units, so that each unit matches the number of processors available and can be processed in parallel. The units are processed individually. However, for many problems such as the solution of a partial differential equation, each point on a square grid $\{(x, y): 1 \leq x \leq n, 1 \leq y \leq n\}$ is updated by linear combination of values at the four adjacent points. In this type of problem, crinkling is preferable, since it maintains the nearest-neighbour connectivity. Both techniques are illustrated in Fig. 9.1 for the case of solving a problem on an 8×8 grid with a 4×4 array of interconnected processors with nearest-neighbour connectivity.

Slicing is straightforward: divide the problem into four units, and call these $A(,, 1, 1)$, $A(,, 1, 2)$, $A(,, 2, 1)$, and $A(,, 2, 2)$, so that each is 4×4 and can be dealt with in parallel. For the case of crinkling: divide the grid into sets of 2×2 blocks, and assign

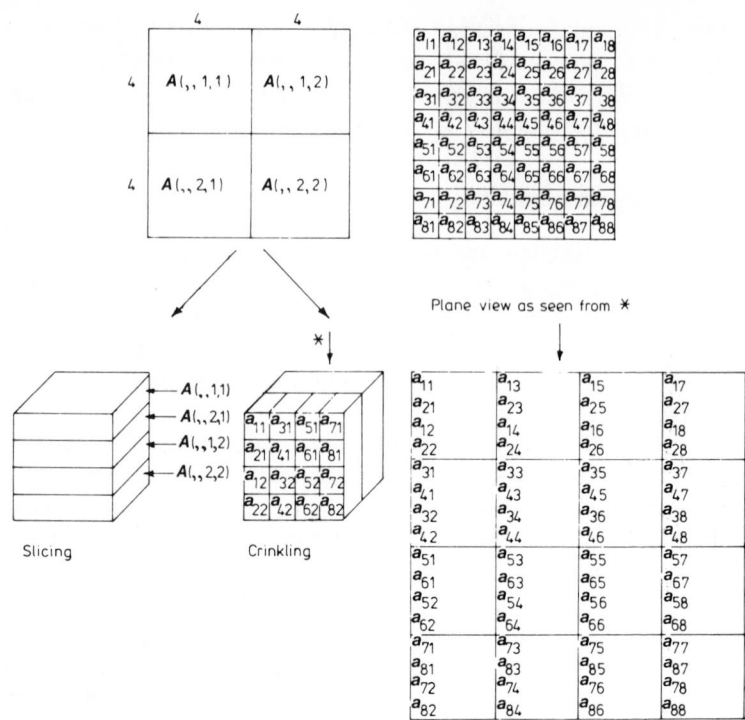

Fig. 9.1 Slicing and crinkling

each set to one processor. The plane view, as seen from ∗ in Fig.
9.1, shows how the nearest-neighbour connectivity at the edges is
preserved. For this model of computation, and assuming nearest-
neighbour connectivity, suppose we wanted to update the matrix
A as follows:

$$B_{ij} = \tfrac{1}{4}(A_{i,j+1} + A_{i-1,j} + A_{i,j-1} + A_{i+1,j}) \quad (i,j = 1,\dots, 8).$$

After crinkling, we can update each of the four partitioned units
simultaneously, and so in four steps complete the update. If, on
the other hand, we use slicing, the nearest-neighbour connectivity
is maintained only for rows and columns up to $n - 1$, with $n = 4$ in
this case. Two special cases would therefore be needed for each
unit, and hence a total of twelve steps!

Depending on the type of problem under consideration, the
above techniques have been used quite extensively on SIMD
architectures: for example in matrix multiplication, eigenvalue

problems, solving sets of linear equations, and in image and speech processing. Further, both slicing and crinkling may be needed in the course of computation. Thus the problem becomes that of changing from one mapping to another. The work of Flanders provides an elegant approach to data organization in general, and is the subject of the final section of this chapter.

Hyafil and Kung introduce the terms *algorithm decomposition* and *problem decomposition,* which are related to the processes of slicing and crinkling (see Heller 1978). They consider the time complexity in terms of the number of parallel steps with fewer processors. Suppose that q_i operations are performed in step i of algorithm 1, with (say) $\beta(N)$ processors; then they assert that this one step becomes $\lceil q_i/\gamma(N) \rceil$ steps in algorithm 2, which restricts the number of processors to $\gamma(N) < \beta(N)$. Brent (1974) has also carried out research in this field, and derived some bounds on the number of steps required with a reduced set of processors available. Of course, recursive doubling (see Section 2.5) is another technique which can be thought of as repeatedly decomposing a problem into smaller versions of the original problem, which then require fewer processors; for example, the product $x_1 x_2 \cdots x_N$ can be expressed as

$$\prod_{i=1}^{N} x_i = \left(\prod_{i=1}^{s} x_i \right)\left(\prod_{i=s+1}^{N} x_i \right),$$

and so on recursively for the subproducts, and hence can be computed in $\lceil \log_2 N \rceil$ steps using $\frac{1}{2}N$ processors.

2.10 Parallel data transforms: a calculus of data organization

The discussion of data mapping in Chapter 1 (Sections 1.5 to 1.8) focuses on routeing networks, which link certain processors directly to others, and specific algorithms whose data communication closely matches those networks. This method of data organization is very effective in practice, and consists of choosing an 'optimal' scheme of mapping and routeing data on to a given network. However, in a number of applications, such as the fast Fourier transform, data interaction can be intricate, and the search for an optimal mapping may prove difficult. A more uniform approach has been proposed by Flanders, which also admits closer analysis. It is called parallel data transforms

(PDTs) and deals with the mapping and routeing of data on processor arrays.

The essential idea is as follows. Rather than consider the representation of data mapping directly onto the routeing network, we consider an alternative representation which is described by a mapping vector. It is designed only for a subset of all possible data mappings, but is sufficiently general to allow the solution of a number of important problems.

(a) *Single data item per processor*

Let us first consider the problem of mapping n data items onto n linearly connected processors, with each processor's memory occupied by a single data item. Clearly there are $n!$ ways in which the data items can be allocated. For example, in the case $n = 8$, we may have the perfect-shuffle mapping, defined as follows.

$$\text{Data item index:} \quad 0 \ 4 \ 1 \ 5 \ 2 \ 6 \ 3 \ 7$$
$$\downarrow \downarrow \downarrow \downarrow \downarrow \downarrow \downarrow \downarrow$$
$$\text{Processor:} \quad 0 \ 1 \ 2 \ 3 \ 4 \ 5 \ 6 \ 7.$$

This can simply be written as [0 4 1 5 2 6 3 7], where data item 0 is placed in processor 0, data item 4 is placed in processor 1, and so on. Such a mapping is referred to as a physical mapping, because it describes the physical arrangement of data items in the processors. It can describe any of the $n!$ mappings, but its specification requires n symbols, which can be cumbersome for large n. The mapping-vector approach, on the other hand, permits only a subset of the $n!$ mappings, but its specification requires only $\log_2 n$ symbols. Consider the above example with eight processors and eight data items. Let the binary value of an index identifying the data item be $b_2 b_1 b_0$. The binary value for the corresponding processor is determined by taking a specific ordering of the bits b_2, b_1, b_0. The ordering $b_1 b_0 b_2$ maps, for example, data item 4 with binary value 100 to 001, i.e. data item 4 is placed in processor 1. The complete bit mapping is as follows.

$$\text{Data item index:} \quad 000 \ 001 \ 010 \ 011 \ 100 \ 101 \ 110 \ 111$$
$$\text{Processor:} \quad 000 \ 010 \ 100 \ 110 \ 001 \ 011 \ 101 \ 111$$
$$\quad\quad (0) \ \ (2) \ \ (4) \ \ (6) \ \ (1) \ \ (3) \ \ (5) \ \ (7)$$

Table 10.1 Mapping vectors and physical mappings for
$n = 8$

Description	Mapping vector	Physical mapping
Identity	(2 1 0)	[0 1 2 3 4 5 6 7]
Bit reversal	(0 1 2)	[0 4 2 6 1 5 3 7]
Perfect shuffle	(1 0 2)	[0 4 1 5 2 6 3 7]
Inverse shuffle	(0 2 1)	[0 2 4 6 1 3 5 7]

The numbers in the last row indicate the processor. We thus
specify the mapping vector as (102), and it corresponds to the
physical mapping [0 4 1 5 2 6 3 7] described above. The mapping vector is based on permutation of the index bits, whereas
physical mapping entails permutation of the actual data items.
The convention is to use parentheses to represent the mapping
vector and square brackets to represent the physical mapping.
We list some mapping vectors and corresponding physical
mappings, for the case $n = 8$, in Table 10.1.

The routeing of data is equivalent to changing the mapping of
data onto processors. Thus we can view the rearrangement of
data items in terms of changes to the mapping vector rather than
in physical space. Furthermore, changes to the mapping vector
can be effected by pairwise exchanges of the elements within it.
Suppose we wish to transform the bit-reversal mapping to the
perfect-shuffle mapping—a problem which may arise, for example, in computing the fast Fourier transform. The initial
mapping (012) becomes (210) by means of a single exchange of
the first and the last binary digits. In general, for a large number
of data items, several pairwise exchanges may be necessary;
however, given any permutation of mapping vector elements,
there is always at least one sequence of pairwise exchanges that
will effect the permutation.

Batcher's bitonic sort algorithm may be implemented using the
perfect shuffle. Here the identity mapping is transformed to the
shuffle mapping. For the case $n = 8$, the mapping (210) becomes
(102), by means of a single left-cyclic shift of the binary indices.
This can be accomplished by two pairwise exchanges in any one
of the following three ways.

$$2\ \underline{1\ 0} \qquad 2\ \underline{1\ 0} \qquad 2\ 1\ \underline{0}$$

$$1\ \underline{2\ 0} \qquad 0\ \underline{1\ 2} \qquad \underline{2\ 0\ 1}$$

$$1\ 0\ 2 \qquad 1\ 0\ 2 \qquad 1\ 0\ 2$$

Using PDTs, the Bitonic sort can be implemented more efficiently than this by a method involving only one pairwise exchange for each step rather than a perfect shuffle; details may be found in Flanders (1982). The pairwise exchanges are referred to as *generators,* and are used to build up transformations between mapping vectors. On the Distributed Array Processor (DAP) such generators and some composite transformations have been implemented using macros.

So far, we have assumed that the mapping is applied to a linear array of processors, with each processor's memory accommodating only a single data item. But PDTs are extended to cover multiple data stored per processor, multidimensional data, and multidimensional processor arrays connected by various routeing networks. These extensions simplify the task of data organization for much more realistic problems than those discussed so far.

(b)   *Multiple data per processor*

Let us consider the mapping of 16 data items on a linear array of four processors. There are a number of ways in which data can be allocated to processors, but we shall consider only the strategies of slicing and crinkling, which were introduced in Section 2.9.

(1)   *Sliced mapping*   The 16 data items are partitioned into four groups, each containing four items, and mapped across the linear array of processors as shown below.

Store address	Processor number			
	0	1	2	3
0	0	1	2	3
1	4	5	6	7
2	8	9	10	11
3	12	13	14	15

This is a two-dimensional analogy of the physical mapping introduced earlier, and can be abbreviated to:

$$[0 \quad 1 \quad 2 \quad 3]$$
$$[4 \quad 5 \quad 6 \quad 7]$$
$$[8 \quad 9 \quad 10 \quad 11]$$
$$[12 \quad 13 \quad 14 \quad 15].$$

Thus data items 0, 4, 8, and 12 are contained in the memory of processor 0; items 1, 5, 9, and 13 are in processor 1, and so on.

(2) *Crinkled mapping* The 16 data items are again partitioned into four groups of four, but the items within each group are mapped under the same processor, as shown below.

Store address	Processor number			
	0	1	2	3
0	0	4	8	12
1	1	5	9	13
2	2	6	10	14
3	3	7	11	15

This mapping may be expressed as follows:

$$[0 \quad 4 \quad 8 \quad 12]$$
$$[1 \quad 5 \quad 9 \quad 13]$$
$$[2 \quad 6 \quad 10 \quad 14]$$
$$[3 \quad 7 \quad 11 \quad 15].$$

The data items 0, 1, 2, and 3 are stored in processor 0; items 4, 5, 6, and 7 are stored in processor 1; and so on.

We shall now derive the mapping vectors. In the previous case, where the number of data items matched the number of processors, the index bits of each data item were permuted to produce a single number indicating the processor in which that particular item was stored. The permutation was defined by a mapping vector. In order to extend the mapping to more than one data item per processor, the natural step is to specify two numbers within the mapping vector: (a) the number of the processor, and (b) the address within it. These two numbers will determine uniquely the location of any data item.

For case (1), the location of each data item is derived from its index $b_3b_2b_1b_0$, by taking the first two digits b_3b_2 to represent the store address, and the last two digits b_1b_0 to represent the processor. For example, 13 with binary value 1101 gives the address index 11, i.e. 3, and the processor index 01, i.e. 1; hence 13 is stored at address 3 in processor 1.

For case (2), the location of each data item, as before, is derived from its index $b_3b_2b_1b_0$, but now the last two digits b_1b_0 represent the store address, and the first two digits b_3b_2 the processor. For the above example, 13 would be stored at address 1 of processor 3. The convention is to drop the b's and represent the mapping vectors as follows.

(a) Slicing: (address: 3 2 | processor: 1 0)
(b) Crinkling: (address: 1 0 | processor: 3 2)

Once again, the transformation between mappings can be effected by invoking a sequence of generators. The transformation from the crinkling to slicing may be performed as follows.

Initial mapping: (address: 1 0 | processor: 3 2)

                  (address: 3 0 | processor: 1 2)

Final mapping: (address: 3 2 | processor: 1 0)

The data movements are slightly more complicated than before, but still very systematic and suited to implementation on a processor array. The inverse transformation, from slicing to crinkling, can be obtained by applying the above sequence of generators in reverse order. The two-parameter specification of the mapping vector can be extended to three parameters, to cover a two-dimensional array of processors. The 'processor' label will have two components: row and column, and together with the 'address' label can specify uniquely the location of any data item. For example, suppose we have in a vector 32 data items which are to be mapped onto a 2×4 processor array p as a matrix of 2 rows and 4 columns, by means of slicing, as shown in Fig. 10.1. Thus processor $p(0, 2)$, for instance, contains items 2, 10, 18, and 26 respectively at addresses 0, 1, 2, and 3.

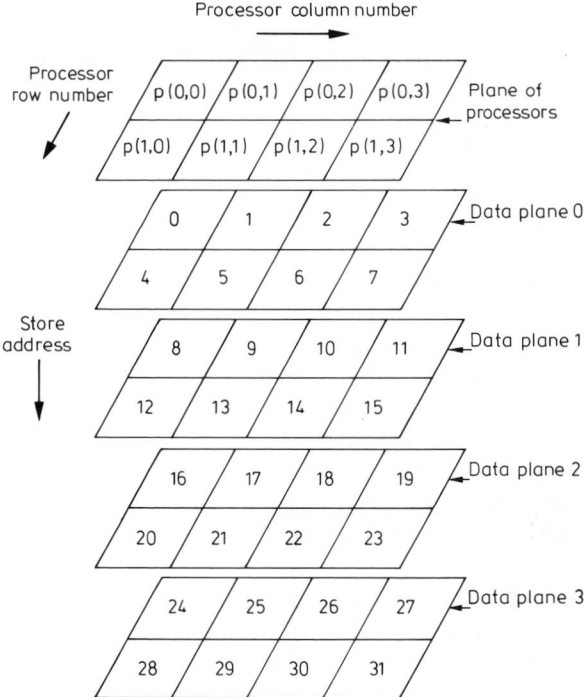

Fig. 10.1 Mapping 32 items on 2×4 processor array

The physical mapping is denoted by

$$\begin{bmatrix} 0 & 1 & 2 & 3 \\ 4 & 5 & 6 & 7 \end{bmatrix}$$
$$\begin{bmatrix} 8 & 9 & 10 & 11 \\ 12 & 13 & 14 & 15 \end{bmatrix}$$
$$\begin{bmatrix} 16 & 17 & 18 & 19 \\ 20 & 21 & 22 & 23 \end{bmatrix}$$
$$\begin{bmatrix} 24 & 25 & 26 & 27 \\ 28 & 29 & 30 & 31 \end{bmatrix},$$

and the mapping vector by

(address: 4 3 | column: 1 0 | row: 2).

Hence,

data item 24, with binary index 11000, is at address 3, in $p(0, 0)$,
data item 18, with binary index 10010, is at address 2, in $p(0, 2)$,
data item 13, with binary index 01101, is at address 1, in $p(1, 1)$,
data item 6, with binary index 00110, is at address 0, in $p(1, 2)$,
etc.

(c) *Multidimensional data*

We now consider the case of a multidimensional array of data. The dimensions of the arrays are assumed to be in powers of two. As in the case of one-dimensional data, where the bits of the data index appear as elements of the mapping vector, the case of multidimensional data is easily extended by specifying indices for each of the dimensions. For example, the two-dimensional 8×16 matrix requires two indices. We label the indices corresponding to the first and second dimensions as the *first* and *second* index respectively. Thus the binary value for the first index requires three bits $b_2 b_1 b_0$, and we denote this by $2_1 1_1 0_1$. Similarly, the second index requires four bits $b_3 b_2 b_1 b_0$, and we denote this by $3_2 2_2 1_2 0_2$. Both these indices appear in the mapping vector, as illustrated in a straightforward example below.

Suppose we wish to map a 2×4 matrix onto a 2×2 processor array using slicing and crinkling. The slicing maps each 2×2 subarray into *one plane*. The crinkling maps each 2×1 subarray into the *memory of one processor*; adjacent subarrays are mapped into adjacent processors. The physical mappings for both are shown in Fig. 10.2. Data items are identified by indices of the form (row index, column index). The index corresponding to the first dimension can be in the range 0 to 1, and therefore requires a single digit, denoted by 0_1. The index corresponding to the second dimension can be in the range 0 to 3, and requires two digits, denoted by $1_2 0_2$. Slicing uses the index bits to form the address as shown below.

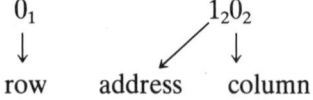

$$0_1 \qquad\qquad 1_2 0_2$$
$$\downarrow \qquad \swarrow \; \downarrow$$
$$\text{row} \quad \text{address} \quad \text{column}$$

Hence the mapping vector for sliced mapping is

$$(\text{address: } 1_2 \mid \text{column: } 0_2 \mid \text{row: } 0_1).$$

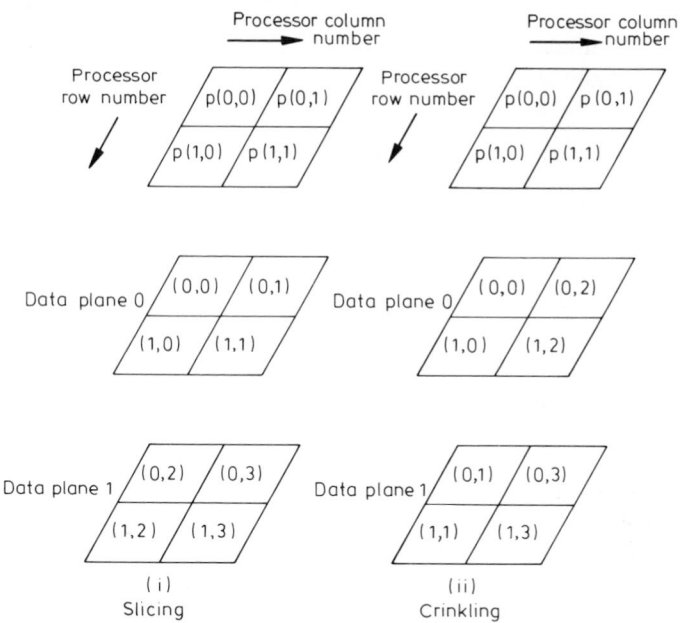

Fig. 10.2 Mapping 2×4 items on 2×2 processor array

The relationship between the mapping vector and the physical mapping can be seen by comparing the mapping of any given element in each case. For example, consider the item in row 1 and column 3 of the 2×4 matrix of data. The first index is 1 (binary 1), thus $0_1 = 1$; and the second index is 3 (binary 11), thus $0_2 = 1$ and $1_2 = 1$. Substituting these values in the mapping vector gives (address: 1 | column: 1 | row: 1). Hence this item is stored at address 1, column 1 and row 1, which is easily verified by checking the position of the item $(1, 3)$ in the physical mapping shown above.

Similarly, it can be shown that the mapping vector corresponding to crinkling is

$$\text{(address: } 0_2 \mid \text{column: } 1_2 \mid \text{row: } 0_1\text{).}$$

As above, the location of any particular data item can be determined from the mapping vector. For example, the item in row 0 and column 1 of the data has $0_1 = 0$, $1_2 = 0$, and $0_2 = 1$; substituting these values into the crinkled mapping vector gives:

(address: 1 | column: 0 | row: 0) indicating that the data item is stored at address 1, column 0 and row 0 of the processor array.

In the above discussion it has been assumed that the individual items are treated as indivisible. However, this restriction can be relaxed by introducing an extra dimension to the data. This is referred to as the *data dimension* and corresponds to indexing along the lengths of data items in units of one or more bits. For single-bit indexing (the principle can be extended to multiple-bit indexing) the data dimension in the mapping vector is specified by $k = \log_2 p$ symbols: $(k-1)_0(k-2)_0 \ldots 0_0$, where individual data items are of length $p = 2^k$ bits. For example, for 8-bit numbers, the data dimension is specified by $2_0 1_0 0_0$; thus the complete mapping vector for crinkling, discussed above, is given by

$$(\text{address: } 0_2 2_0 1_0 0_0 \mid \text{column: } 1_2 \mid \text{row: } 0_1).$$

To locate bit 3 of the data item $(0, 1)$, for example, we substitute the values

$$2_0 = 0, \; 1_0 = 1, \; 0_0 = 1 \quad \text{(since bit-number = 3)}$$
$$0_1 = 0 \quad \text{(since first index = 0)}$$
$$1_2 = 0, \; 0_2 = 1 \quad \text{(since second index = 1)}$$

into the mapping vector to give (address: 1011 column: 0 row: 0). Hence, bit 3 is located at bit-plane 11 (binary 1011) of the processor in column 0 and row 0. Clearly the use of the data dimension is only necessary if, in the course of computation, individual data items are broken down into subparts. Otherwise their positions in the mapping vector remain unaltered.

Data routeing is an integral part of parallel computation. The use of PDTs allows efficient mapping and routeing of data, enabling us to keep track of the data even in complex situations where the mapping of data changes dynamically throughout the algorithm.

The implementation of PDTs on the DAP allows transformations between mappings to be specified explicitly as sequences of generators; the code to implement the physical transformations corresponding to the generators is produced by PDT system software from the specification of the generator sequence. Later developments permit the user simply to specify two mapping vectors and the PDT software determines an efficient sequence of generators automatically; the determination of the generator sequence may take place either at compile-time or at run-time, depending on when the details of the mapping vectors are known.

As well as being useful in implementing one-off data routeing operations, mapping vectors have important properties with respect to the pairing of data for parallel operations. PDTs are therefore useful in the mapping of complete algorithms on to processor arrays; this has been demonstrated on, for example, Batcher's bitonic sort and the fast Fourier transform.

PDTs are further extended as follows:

(a)   the class of mappings handled is increased by allowing the inverse of index bits to be used when forming the location of the data;

(b)   the theory is applicable to processor arrays with the processors connected by networks other than nearest-neighbour, e.g. perfect shuffle and binary hypercube.

For problems involving multidimensional data and multidimensional arrangements of processors, the mapping vector nevertheless retains the clarity of its single dimension, with a reduction, by a factor of $n/\log_2 n$, in the amount of data needed for its specification. The process has been demonstrated on the DAP, and has enhanced our understanding of the effects of different routeing networks. A very readable introduction to parallel data transforms, containing a number of examples, is provided by Flanders and Parkinson (1988), and a more formal analysis by Flanders (1982).

3 Parallel sorting algorithms

3.1 Introduction

The sorting of a sequence of numbers into increasing order, and locating a particular item in a file of data, constitutes the 'nuts and bolts' of computer science. Sorting assists in the administration of files and records, and is closely related to searching. We will explore this absorbing field, albeit briefly, to gain insight into the ways that parallelism may be incorporated into well-established techniques. This will provide an intellectual exercise in demonstrating concurrency, and is thus guaranteed to prepare the reader for the more rigorous demands imposed in subsequent chapters by the eigenvalue and singular-value problems.

Extensive research into sorting techniques has been carried out during the last few years, and a large volume of literature is available—notably Knuth's book (1973), a more recent one by Akl (1985), which provides an introduction to parallel sorting methods, and a paper by Bitton *et al.* (1984) which presents a taxonomy of parallel sorting. Much of the attention has focused on restructuring well-known serial techniques such as quicksort, odd–even transposition sort, and Batcher's bitonic sort in order to make them amenable to parallelism. This chapter provides a synthesis of recent work in parallel sorting techniques, and introduces the idea of complete parallel sorting schemes.

We shall consider *parallel internal sorting methods* where the items to be sorted can be accommodated within the total memory size of the parallel system (considered as concatenation of the processors' local memory). For *parallel external sorting methods*—where the items exceed the total memory size and part of the file may reside on magnetic tape or disk—the reader is referred to a paper by Even (1974). These strategies are equivalent to slicing and crinkling (Chapter 2, Sections 2.9 and 2.10). The algorithms are not fundamentally different from those for internal sorting methods, to which we now turn our attention.

The numbers $a_1, \ldots, a_n$, initially in locations $1, \ldots, n$, are to be repositioned so that the smallest occupies location 1, the second smallest location 2, and so on. There are several possible strategies: *insertion sort*, where each item is inserted into the correct position relative to previously sorted items; *selection sort*, where the smallest item is selected and removed from the list, then the next smallest, and so on; *enumeration sort*, where each item is compared with the others and the number of smaller keys indicates its position in the sorted list; *comparison–exchange sort*, where the current occupants x_i and x_j of two locations i and j $(i < j)$ are interchanged if $x_i > x_j$. We will concentrate particularly on parallel methods based on comparison–exchanges; for a detailed discussion of the others the reader is referred to Knuth (1973). On a serial machine, each comparison–exchange must be carried out consecutively; parallel algorithms differ in that, in a single step, many comparison–exchanges may be performed simultaneously, subject to the condition that no two involve a common location. The general aim in constructing a parallel sorting algorithm is to minimize the *delay* (Knuth's term) or the number of such steps or *parallel comparisons*.

A number of important results have appeared recently, and some of the main ones are summarized here. A serial algorithm based on comparison–exchanges necessarily requires at least $O(n \log_2 n)$ comparisons to sort n numbers, because, if there are N comparisons, then these can distinguish between at most 2^N different permutations of $1, \ldots, n$, and there are $n!$ altogether: thus, if all arrangements can be sorted, then $N \geq \log_2 n!$, which is asymptotically $n \log_2 n$ by Stirling's formula $n! \sim (n/e)^n (2\pi n)^{\frac{1}{2}}$. If n comparisons are carried out simultaneously at each stage, then clearly the lower bound on the number of parallel comparisons (or delay) is $O(\log_2 n)$. However, it does not seem possible to achieve this lower bound by restructuring one of the well-known $O(n \log_2 n)$ serial algorithms (for example, the two-way merge sort), primarily because of lack of parallelism towards the end of the sorting process. On the other hand, it is possible to parallelize fully a serial sorting method requiring $O(n^2)$ comparisons so that the optimal delay of $O(n)$ is achieved: the odd–even transposition sort and the parallel tree sort (discussed later in this chapter) both use $O(n)$ processors to sort n numbers in $O(n)$ steps.

We next consider sorting based on special-purpose networks consisting of a number of processors, usually called *comparators,* interconnected in such a way as to implement directly a parallel sorting algorithm. Any one comparator is capable of performing comparison–exchange on pairs of numbers. This approach has proved effective, and indeed one of the first results in parallel sorting was due to Batcher (1968), who devised a technique based on sorting networks which took $\frac{1}{2}(\log_2 n)(\log_2 n + 1)$ parallel comparisons to sort n numbers (or keys). Stone (1971) showed that Batcher's sort, when implemented on the *perfect-shuffle* interconnection (see Chapter 1, Section 1.7), requires as few as $\frac{1}{2}n$ comparators. These schemes clearly represent a great improvement over the $O(n)$ complexity of the algorithms mentioned above and are discussed in detail in the following sections. In order to improve the bounds still further, some authors (Ajtai *et al.* 1983) have carried out theoretical studies using networks based on complex graph construction, which achieve the lower bound of $O(\log_2 n)$ comparisons to sort n numbers. These may, however, involve an unacceptably high proportionality constant in the expression for the lower bound, and are thought not to be suitable for implementation (Bitton *et al.* 1984).

Another highly parallel algorithm, referred to as *siftsort,* is asserted to take $O(\log_2^2 n)$ steps to sort n items of data on fewer than $n \log_2 n$ processors. Although a comparison–exchange algorithm, it separates the comparisons from the exchanges, with an intermediate logical step of selectively annihilating—i.e. sifting— an array representing the results of the comparisons. The algorithm proceeds by successive sweeps, with each sweep consisting of three stages: comparison, sifting, and exchange. Unlike Batcher's sort, the method is robust against random errors occurring in the sorting process. For further details, the reader is referred to Bentin and Modi (1988).

Recently, theoretical models of parallel processors with shared memory (see Chapter 1, Fig. 2.1) have been proposed, which can sort n numbers in $O(\log_2 n)$ comparisons, but the algorithms based on enumeration require $O(n^2)$ processors. The idea is a simple one. Sorting is performed by computing in parallel the rank of each number by comparing it with all the others (for a pair of numbers x and y the binary bit for x is 1 if $x > y$, otherwise 0). With a shared-memory system this is possible because any processor may access any location in the memory. Thus, to sort a sequence of n numbers, each is compared simultaneously with all the others in a single step using $n(n-1)$ comparators. The binary set obtained in each case is fed into a counter which counts the number of 1's, and, if this is p, then the position of x in the sorted list is at location $p+1$. It takes $\log_2 n$ steps to compute the rank using a parallel counter. The numbers are then routed to the correct position through a switching network (for further details, see Muller and Preparata 1975). The idea of combining enumeration with parallel merge has also been investigated by Preparata (1978).

Sorting may be viewed as a *graph-theoretical problem*, which is particularly relevant in the parallel case. The relationship can be expressed as follows. Let V be a set of n elements in arbitrary order, K_n the complete graph on V, and T a transitive orientation of K_n, i.e. T is a *linear order* on V. We know V, and our aim is to find T by making a minimum number of probes: a *probe* of the edge $v_1 v_2$ of K_n reveals whether $v_1 v_2$ or $v_2 v_1$ is an arc of T. The result of probing can be represented by an acyclic digraph on V, containing arcs of T discovered by probing, together with those arcs of T implied by the transitivity of T. It can be shown that the minimum number $p(n)$ of probes needed in the worst case is given exactly by

$$p(n) = kn - 2^k + 1 \quad \text{for } 2^{k-1} \leq n < 2^k$$

($k = 1, 2, \ldots$). Thus any sorting algorithm will need at least $n(\log_2 n - 1) + 1$ probes in the worst case. On a parallel machine, several probes can be made at any one step. For a detailed description of this approach see Bollobás and Hell (1985).

In Section 3.8, we provide a resumé of the recent paper by Davies and Modi (1986*b*) which introduces the idea of *complete* sorting schemes. These are derived from algebraic problems and are not competitive with the methods mentioned above. However, they may be of independent interest; for instance, the sorting of eigenvalues could be incorporated in the parallel Jacobi method.

Present-day parallel computers receive input sequentially, which can cause serious delay. Indeed, if a machine receives a single datum per unit time, then, for n items, n units of time are required to load the data. It is quite possible, however, to envisage a parallel input and output facility in the not too distant future. Rather than enter into debate over whether or not such parameters should be included in the analysis of algorithms, we shall assume throughout the rest of this chapter that data are initially distributed amongst the processors. This assumption is quite reasonable, if the numbers are themselves the result of some parallel computation.

3.2 Batcher's bitonic sort

A sequence $(z_1, ..., z_n)$ is *bitonic* if, after some cyclic shift, it consists of an ascending sequence followed by a descending sequence. For example (3, 5, 6, 6, 4, 2, 1, 3) is bitonic because it can be shifted to (1, 3, 3, 5, 6, 6, 4, 2). (Earlier authors, such as Knuth (1973), did not allow the cyclic shift and considered only what one might term *strictly bitonic* sequences.)

The basis of the parallel version of Batcher's sort is the easily verified fact that if $(z_1, ..., z_{2n})$ is a bitonic sequence of even length, and in one parallel step we perform n comparison-exchanges so as to form

$$\min(z_1, z_{n+1}), ..., \min(z_n, z_{2n}), \max(z_1, z_{n+1}), ..., \max(z_n, z_{2n}),$$

then the two halves of this are bitonic and contain the n smallest and n largest elements respectively of the original sequence. Given any bitonic sequence, and assuming for simplicity that it has length $n = 2^k$, it is clear that if the above process is applied first to the whole sequence, then to each half, then to each half of these halves, and so on, then in k parallel steps the whole sequence will be sorted.

In Fig. 2.1, we take $k = 3$ and $z = (3, 4, 7, 9, 6, 4, 2, 0)$. The horizontal lines correspond to locations, and the pairs of elements undergoing a comparison–exchange are indicated by the vertical arrows; the direction of the arrow (always downwards in this case) indicates, by the position of the head, the location in which the largest element is to be placed. The intermediate result is shown after each parallel step.

The above example started with a strictly bitonic sequence; the process works equally well for a nonstrictly bitonic sequence. With $k = 3$, for example, we have $z = (7, 9, 6, 4, 2, 0, 3, 4)$, and

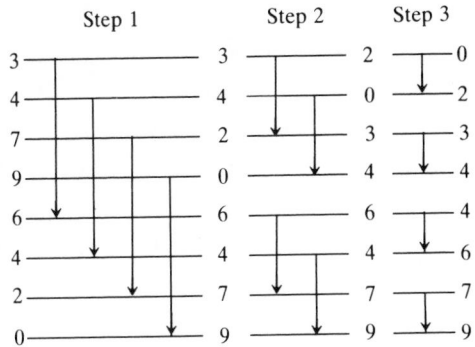

Fig. 2.1 Comparison–exchange scheme for $n = 8$

the intermediate results after each parallel step are as follows:

Step 1

$(7, 9, 6, 4, 2, 0, 3, 4) \rightarrow (2, 0, 3, 4, 7, 9, 6, 4)$

Step 2   Step 3

$\rightarrow (2, 0, 3, 4, 6, 4, 7, 9)$ $\rightarrow (0, 2, 3, 4, 4, 6, 7, 9)$.

Now suppose that $z_1, \ldots, z_n$ is an arbitrary sequence, again with $n = 2^k$, and not necessarily bitonic. By means of one parallel step, we get (say)

$$z_1 \leqslant z_2, \qquad z_3 \geqslant z_4, \qquad z_5 \leqslant z_6, \qquad z_7 \geqslant z_8, \ldots .$$

Each of the four-element subsequences (z_1, z_2, z_3, z_4), $(z_5, z_6, z_7, z_8), \ldots$, is bitonic and, by the previously described method, can be sorted in two parallel steps. For subsequences of odd order, we apply the scheme 'with arrows reversed', so that the result satisfies

$$z_1 \leqslant z_2 \leqslant z_3 \leqslant z_4, \qquad z_5 \geqslant z_6 \geqslant z_7 \geqslant z_8, \ldots .$$

Each of the eight-element subsequences, such as $(z_1, \ldots, z_8)$, is bitonic and can be sorted in three parallel steps. Continuing in this way, the whole sequence is sorted in $1 + 2 + \cdots + k = \frac{1}{2}k(k + 1) = \frac{1}{2}(\log_2 n)(\log_2 n + 1)$ parallel steps. Cases when n is not a power of 2 can be dealt with by a minor adaptation of the method, or by adding dummy elements to the sequence; the number of parallel steps required remains approximately $\frac{1}{2}(\log_2 n)^2$.

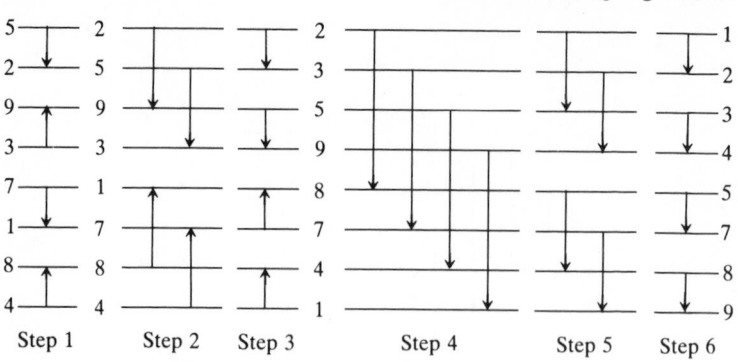

Fig. 2.2   Batcher's bitonic sort for $n = 8$

In Fig. 2.2 the Batcher sort is presented for the case $n = 8$, with $k = 3$ and $z = (5, 2, 9, 3, 7, 1, 8, 4)$. Here the result appears in $1 + 2 + 3 = 6$ parallel comparisons.

3.3   Bitonic sort using the perfect shuffle

Batcher's bitonic sort can be implemented efficiently using the perfect-shuffle interconnection described in Chapter 1 (Section 1.7). The perfect shuffle has certain special properties which make it particularly well adapted to sorting (more details can be found in Akl 1985).

Property 1   Under the perfect-shuffle mapping of n objects, where $n = 2^k$, the items return to their original position after k shuffles.

The second property pertains to the data items placed in adjacent pairs of processors: (P_0, P_1), (P_2, P_3),.... The binary indices, for any pair, differ in the rightmost bit only.

Property 2   Consider the data items in pairs of processors whose indices have binary representations $b_{k-1}b_{k-2}\ldots b_1b_0$ and $b'_{k-1}b'_{k-2}\ldots b'_1b'_0$, which differ only in position $k - r$ where $1 \leq r \leq k$. After r shuffles, these items are located in adjacent processors. (This is more easily seen by considering the cyclic shift of binary representation.)

Consider the bitonic sort for $n = 8$, Fig. 2.2, with the eight input lines labelled as 000, 001, 010, 011, 100, 101, 110, 111.

Any pair of labels, corresponding to horizontal lines connected by an arrow, differs in *exactly one* binary place. Let the ith label have the binary representation $b_{k-1}b_{k-2}\cdots b_1 b_0$, where $n = 2^k$. The *pivot* bits for successive steps of the bitonic sort are as follows:

Step 1 b_0

Step 2 b_1, b_0

Step 3 b_2, b_1, b_0

$\vdots$

Step k $b_{k-1}, \ldots, b_0$.

The sequence $(b_{k-1}, \ldots, b_0)$ consists of k subsequences; at step s, the subsequence has length s consisting of the bits $b_{s-1}, \ldots, b_0$. The properties of the perfect shuffle clearly indicate that we can sort the numbers in the list $L = (x_0, \ldots, x_7)$ using the connection shown in Fig. 3.1, which is derived from the directed graph D_3 of Fig. 7.3 in Chapter 1, for the case $n = 8$.

After one shuffle, each comparator receives two numbers from L, where the indices differ in the leftmost bits of their binary representations; each subsequent shuffle moves the differing bits one position to the right. To implement the bitonic sort on the perfect-shuffle interconnection, L must be shuffled a number of times before each step to ensure that the pivot bit at step s is b_{s-1}. The following two steps realize this objective.

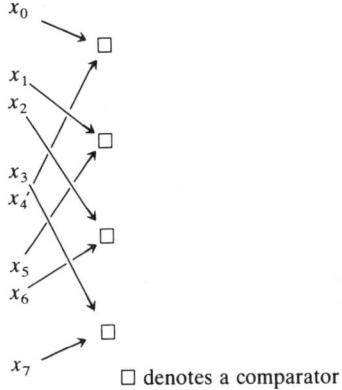

□ denotes a comparator

Fig. 3.1 Data fed into comparators after one shuffle

(i)   For each of the pivots, a comparison–exchange step is followed by a shuffle on L until b_0 is reached.

(ii)   Since there are s pivots associated with step s, the sequence must be shuffled $k - s + 1$ times before step s, which is followed by s comparison–exchange steps and $s - 1$ shuffle steps.

The perfect-shuffle interconnection is shown in Fig. 3.2(i). The stages involved for the case $n = 8$ are shown in Fig. 3.2(ii), with an example of $L_1 = (3, 9, 2, 6, 1, 0, 4, 8)$ indicating how the data propagate through the network in Fig. 3.2(iii). (In this case the first two shuffles are redundant.)

The implementation requires only $\frac{1}{2}n$ processors, or comparators, at the expense of additional steps required to shuffle the data (without performing comparison–exchanges). The number of comparison–exchanges is the same as for Batcher's sort: $\frac{1}{2}(\log_2 n)(\log_2 n + 1)$. The solution idea was first proposed by Stone (1971). Siegel (1977) has since considered several other topologies, including (i) the hypercube and (ii) plus–minus 2^i networks, and shown that the data exchanges required by the bitonic sort can be efficiently performed by these networks. In

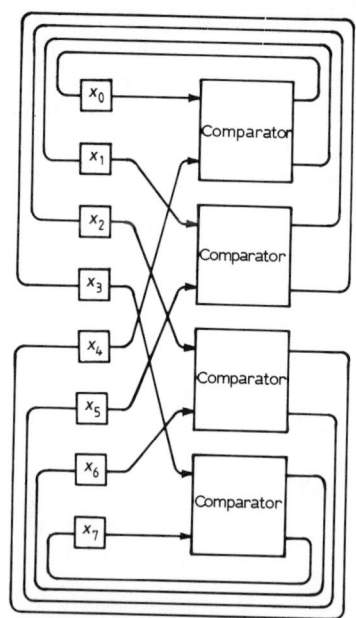

(i) Perfect-shuffle interconnection

Fig. 3.2   Bitonic sort with perfect-shuffle interconnection

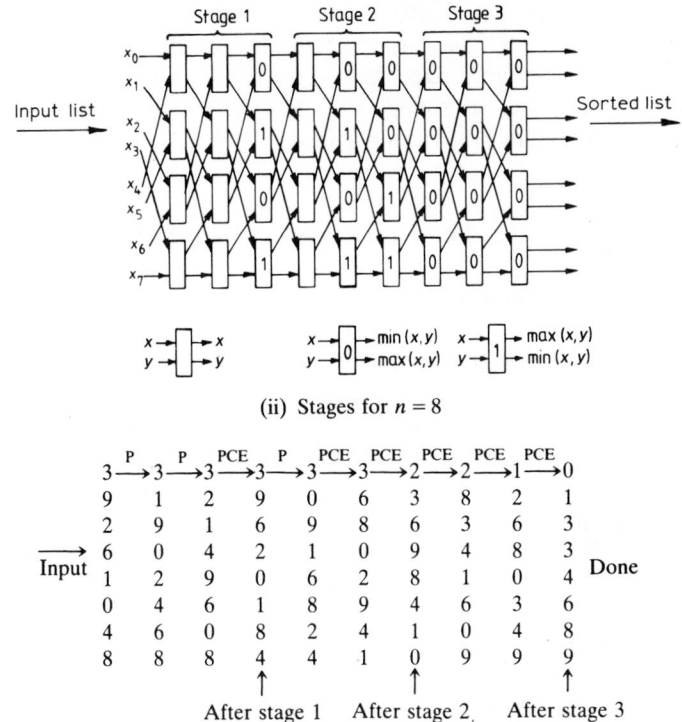

(ii) Stages for $n = 8$

P	P	PCE	P	PCE	PCE	PCE	PCE	PCE	
$3 \rightarrow$	$3 \rightarrow$	$3 \rightarrow$	$3 \rightarrow$	$3 \rightarrow$	$3 \rightarrow$	$2 \rightarrow$	$2 \rightarrow$	$1 \rightarrow$	0
9	1	2	9	0	6	3	8	2	1
2	9	1	6	9	8	6	3	6	3
6	0	4	2	1	0	9	4	8	3
1	2	9	0	6	2	8	1	0	4
0	4	6	1	8	9	4	6	3	6
4	6	0	8	2	4	1	0	4	8
8	8	8	4	4	1	0	9	9	9

$\rightarrow$ Input (rows 3–4), Done (right side)

After stage 1   After stage 2   After stage 3

P = shuffle
PCE = shuffle followed by comparison–exchange

(iii) Sorting the list 3, 9, 2, 6, 1, 0, 4, 8

Fig. 3.2 Bitonic sort with perfect-shuffle interconnection

particular, the topology with cube-connected cycle (CCC) can economically embed both the cube and the perfect shuffle, and requires only three communicating ports per processor. On the DAP, which has a lattice connectivity, the theoretical time, based on parallel data transforms, for sorting n integers each of length b bits is

$$2^{b-12}[0.35(\log_2 n)(1 + \log_2 n) + 148.6]n \qquad (\mu\text{sec}).$$

The term proportional to $n(\log_2 n)(1 + \log_2 n)$ is the contribution from comparison–exchanges, and the term proportional to n is that from remapping data. Thus, for small n, the time is approximately proportional to n, growing asymptotically as

$n(\log_2 n)^2$ for large n. Using the above formula with $n = 2^{18}$ and $b = 5$, the predicted time is 550 msec for sorting a quarter of a million 32-bit integers. Details may be found in Flanders (1982). Further literature on sorting via networks includes papers by Nassimi and Sahni (1982), and Pease (1977).

3.4 Bubble sort and odd–even transposition sort

The bubble sort is a serial algorithm which sorts by comparison–exchange. For $i < j$, we shall denote by (i, j) the comparison–exchange in which the numbers occupying locations i and j are interchanged if the former is larger.

At the first stage, the largest element 'bubbles up' to its correct location by means of the comparison–exchanges

$$(1, 2), (2, 3),\ldots, (n - 1, n);$$

at the second stage, the process is repeated on the locations $1,\ldots, n - 1$ in order to relocate the second largest, and so on. The number of steps is high, namely $\frac{1}{2}n(n - 1)$ if the sequence is initially in the reverse order. The scheme, however, is conceptually simple and uses only comparison–exchanges between adjacent locations. In the most natural parallelization, each stage would begin as soon as the required elements become available; the diagram for $n = 7$ is shown in Fig. 4.1. The horizontal lines correspond to locations, and the pairs of elements undergoing comparison–exchange are indicated by vertical arrows; the direction of the arrow indicates the location in which the larger element is to be placed. The number of parallel steps is $2n - 3$.

The odd–even transposition sort differs only in that a maximum of comparison–exchanges (without common locations) is carried out at each stage, and only n parallel steps are needed. The diagram for $n = 7$ is shown in Fig. 4.2. The process begins with an odd phase consisting of the comparison–exchanges

$$(1, 2), \quad (3, 4), \quad \ldots \quad,$$

followed by the even phase

$$(2, 3), \quad (4, 5), \quad \ldots \quad,$$

and these then alternate.

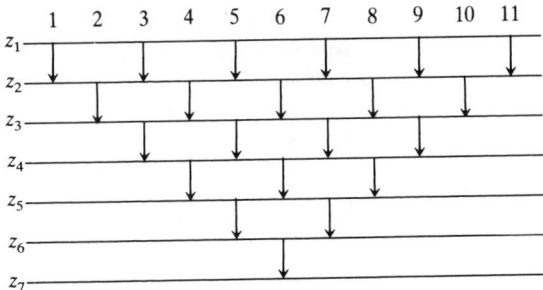

Fig. 4.1 Parallel bubble sort for $z_1,...,z_7$

For a partially ordered list, the method can be rendered more efficient by keeping track of whether or not, at any odd–even phase, there have actually been any comparison–exchanges. If there have not, then clearly the list is sorted.

Example: We take $n = 7$, $z = (3, 4, 1, 2, 5, 6, 7)$; using Fig. 4.2, the intermediate results after each stage are as follows (Table 4.1).

In this case sorting is accomplished in only four stages because, at the 5th and 6th, no comparison–exchanges have taken place. However, for a random list, it would generally take seven stages. Implementation of the method on a linear connection of processors is discussed in Chapter 1, Section 1.5.

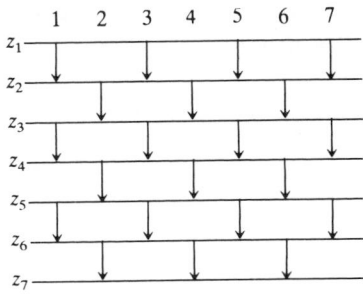

Fig. 4.2 The odd-even transposition sort for $z_1,...,z_7$

Table 4.1 The odd–
even transposition sort
for (3, 4, 1, 2, 5, 6, 7)

3⎞	3	3⎞	1	1
4⎠	4⎞	1⎠	3⎞	2
1⎞	1⎠	4⎞	2⎠	3
2⎠	2⎞	2⎠	4⎞	4
5⎞	5⎠	5⎞	5⎠	5
6⎠	6⎞	6⎠	6⎞	6
7	7⎠	7	7⎠	7

3.5 Tree sort

The tree sort is based on repeated comparisons of pairs of
elements but, unlike the preceding methods, is not exactly a
comparison–exchange method. A list of n numbers is sorted by
finding the minimum and removing it, by finding the next
minimum and removing that, and so on. The scheme is particularly suited to a binary interconnection of processors (which is
discussed in Chapter 1, Section 1.7), and is illustrated here for
the case where there are eight numbers $z_1, ..., z_8$ to be sorted.
Seven processors $P_1, ..., P_7$ are used, with the connections indicated in Fig. 5.1. We start by assigning one number to each
'leaf', as shown.

Each processor at a given level k ($k = 0, 1, 2$ in Fig. 5.1) is
directly connected to two *daughter locations* (processors or
'leaves') at level $k + 1$, and after any given step each location is
either *empty* or *occupied* (storing one number from the list);
initially, all processor locations are empty.

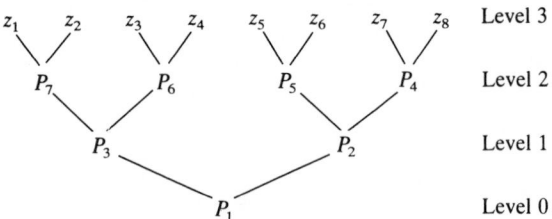

Fig. 5.1   Parallel tree sort for $z_1, ..., z_8$

At each step, (i) every empty processor receives any number(s) occupying its daughter locations and, if there are two, compares them, retains the smaller, and returns the larger to its original location, (ii) if the root processor P_1 is occupied by a number, then this is removed. All processors in the entire network operate simultaneously.

Since all the numbers eventually reach P_1, and in increasing order, sorting is accomplished. After the minimum reaches the root, in $\log_2 n$ steps if n is a power of 2, new numbers reach the root at every second step. The sorting of n numbers takes

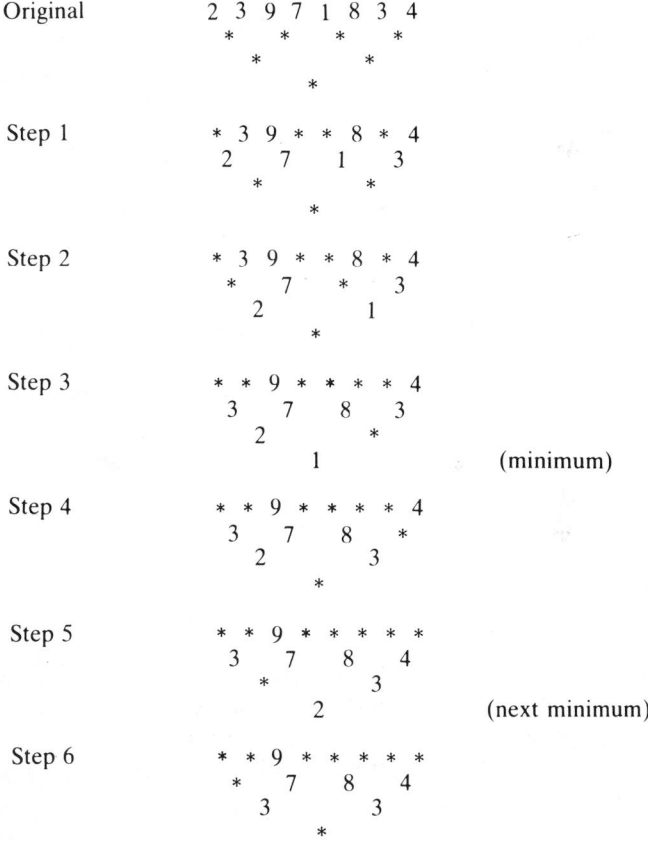

Fig. 5.2 First 6 steps of parallel tree sort for $(2, 3, 9, 7, 1, 8, 3, 4)$

$\lceil \log_2 n \rceil + 2n - 1$ parallel steps, using $O(n)$ processors. The method is illustrated with an example in Fig. 5.2, with $n = 8$, and $z = (2, 3, 9, 7, 1, 8, 3, 4)$; for further discussion, see Bentley and Kung (1979).

3.6 Quicksort

Hitherto, the sequence of comparisons has been prearranged. In the quicksort method, which is due to Hoare (1962), and which is sometimes referred to as the *partition–exchange sort*, the result of each comparison determines which elements are compared next. The order in which the comparisons take place is therefore not predetermined, and consequently it would appear difficult to specify the parallelism schematically and a priori. However, a variant of the method is adaptable to a bit-organized model of parallel computation (Chapter 1, Section 1.3). We discuss those opportunities for parallelism that arise, after introducing the method.

Given the numerical data, or keys $x_1, \ldots, x_n$, the quicksort method ranks the keys with respect to the first key x_1, so that the keys to the left of x_1 are smaller than x_1 and those to the right of x_1 are larger. Hence the location of x_1 in the final list is determined. The process is then repeated recursively on the sublists on either side of x_1, for each sublist of the sublists, and so on. The algorithm is specified using flow-charts consisting of three routines (Figs 6.1–6.3).

Quicksort(n), Fig. 6.1, sorts in order of magnitude the keys $x_1, \ldots, x_n$ into locations $1, \ldots, n$, i.e. it constructs a list $(x(1), \ldots, x(n))$, which is a permutation of $(x_1, \ldots, x_n)$, such that $x(1) \le \cdots \le x(n)$.

Subroutine partition($\bar{S}$), Fig. 6.2, partitions a subset $\bar{S}$ of $\{1, \ldots, n\}$ into maximal subintervals: $\bar{S} = \bigcup_{\lambda=1}^{k} S_\lambda$, where $S_\lambda := \{l \in \bar{S} : i_\lambda \le l \le j_\lambda\}$.

The subroutines subquicksort(i_λ, j_λ) ($\lambda = 1, \ldots, k$), Fig. 6.3, may be run in parallel. Subquicksort(i_λ, j_λ) sorts datum x_{i_λ} into its correct location k_λ (say). Each run of subquicksort(i_λ, j_λ) corresponds to a partitioning of the integer interval $\{i_\lambda, \ldots, j_\lambda\}$ into three subintervals $\{k_\lambda\}$, $\{i_\lambda, \ldots, k_\lambda - 1\}$, and $\{k_\lambda + 1, \ldots, j_\lambda\}$, and thus doubles the potential parallelism successively with each cycle of quicksort when the latter two subintervals are nontrivial for $\lambda = 1, \ldots, k$.

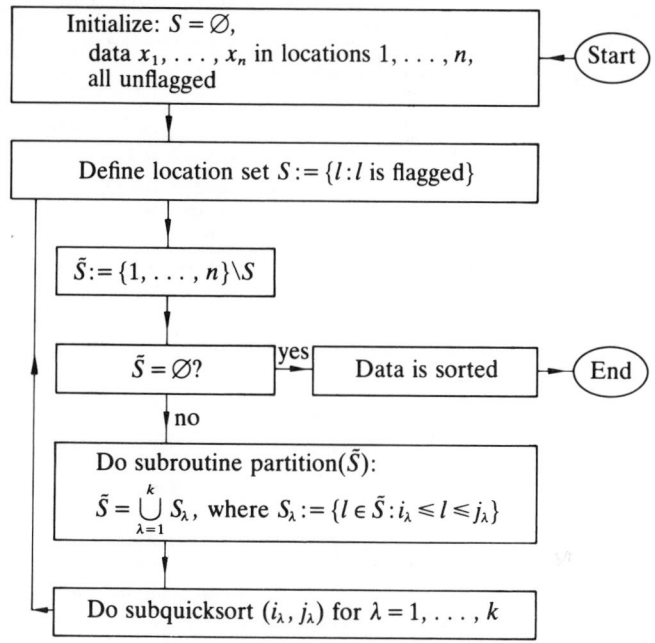

Fig. 6.1 Quicksort(n)

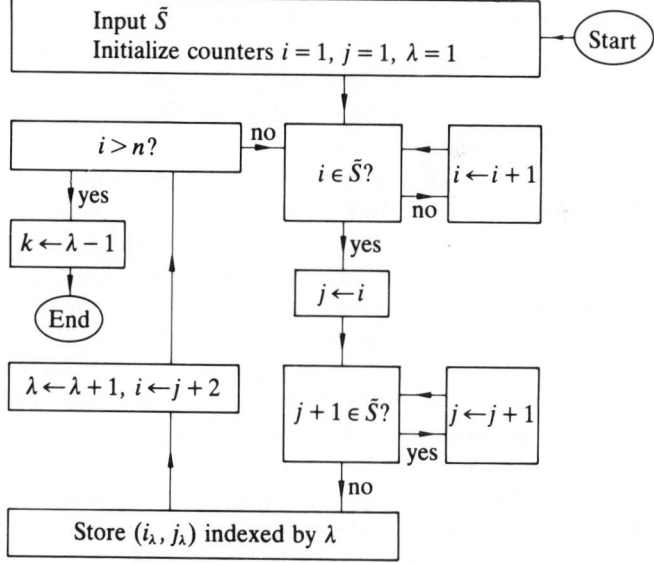

Fig. 6.2 Subroutine partition($\tilde{S}$)

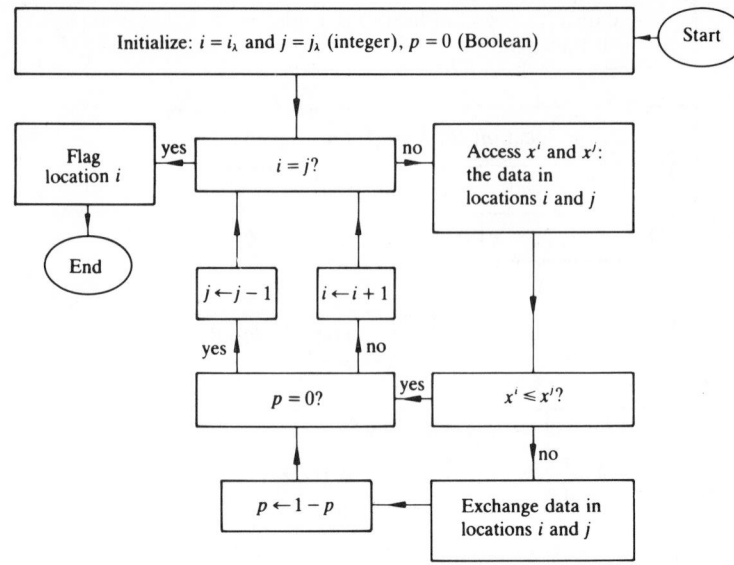

Fig. 6.3 Subquicksort(i_λ, j_λ)

Example Suppose we take $(x_1, \ldots, x_9) = (4, 5, 6, 3, 4, 5, 6, 7, 8)$. Table 6.1 lists the contents of the data registers and the state of the counters immediately after the mth parallel data exchange ($m = 0, \ldots, 9$). Asterisks indicate data in flagged (i.e. sorted) locations. The values of i ($=j$) and p for these locations are omitted after one increment of m.

Unlike the odd–even transposition sort, quicksort does not benefit from ordered keys. In fact, a monotonically increasing or decreasing sequence gives the worst-case behaviour, requiring $\frac{1}{2}n(n + 1)$ comparisons. For random data, however, quicksort achieves on average the lower bound of $n \log_2 n$ comparisons (Knuth 1973), and for serial computation the method is very efficient, requiring only modest amounts of data movement for internal sorting. It has received much attention, and several refinements have been proposed (see, for example, Singleton 1969). One of them is the use of a different sorting algorithm when the sublists are small, such as the bubble sort; another is to rank all keys with respect to the median (see below) in order to obtain a reasonably balanced tree.

Table 6.1 Quicksort applied to $(4, 5, 6, 3, 4, 5, 6, 7, 8)$

m	x^1	x^2	x^3	x^4	x^5	x^6	x^7	x^8	x^9	i	j	p
0	4	5	6	3	4	5	6	7	8	1	9	0
1	3	5	6	4	4	5	6	7	8	1	4	0
2	3	4	6	5	4	5	6	7	8	2	4	1
3	3	4*	6	5	4	5	6	7	8	2	2	0
4	3*	4*	5	5	4	6	6	7	8	1 3	1 6	0 0
5	3*	4*	5	5	4	6*	6	7	8	6	6	1
6	3*	4*	4	5	5	6*	6*	7	8	3 7	5 7	0 0
7	3*	4*	4	5	5*	6*	6*	7*	8	5 8	5 8	1 0
8	3*	4*	4*	5	5*	6*	6*	7*	8*	3 9	3 9	0 0
9	3*	4*	4*	5*	5*	6*	6*	7*	8*	4	4	0

In the following discussion, we shall consider the parallelization of the method for the latter case where the median is not assumed to be the first key. The core of the method consists of the following two steps.

Step 1 Select the key representing the median, and compare it with the others to determine those larger and smaller.

Step 2 Partition the keys into two sublists:

{keys larger than median} {keys smaller than median}.

The use of pointers i and j in the serial algorithm, described previously, disguises the natural parallelism of the method. The above simple description removes the use of pointers completely. It can be seen that Step 1 is well suited to parallelization because, once the median is chosen, all keys are simultaneously compared with it, and a logical mask may be set to indicate the result. In Step 2, sublists can be processed rapidly by means of a parallel switching network that simultaneously routes the keys to their destination.

A careful choice of median is important in order to partition

keys	binary	1st stage	2nd stage	3rd stage	4th stage	result
13	1101	0111	0111	0100	0100	4
7	0111	0110	0110	0111	0110	6
13	1101	0100	0100	0110	0111	7
6	0110	1101	1101	1101	1100	12
4	0100	1101	1101	1101	1101	13
12	1100	1100	1100	1100	1101	13
	↑	↑	↑	↑		

Fig. 6.4 Quicksort based on most significant bit

the keys into two roughly equal sublists. One possibility, which is easily adaptable to a bit-organized array of processors, is to select the most significant bit in the key field as an indicator. If the keys are uniformly distributed, we can expect an equal number of 1's and 0's at any bit location. Thus, through this approach, Step 1 involves no comparisons. Let us consider, as an example, the keys 13, 7, 13, 6, 4, 12 represented in binary, as shown in Fig. 6.4.

The vertical arrow indicates the significant digit under consideration. At the first stage the keys are classified: those with the most significant bit '1' belong to one sublist and those with '0' belong to the other. Step 2 involves a compression operation on the keys so as to form the two sublists: $(7, 6, 4)$ and $(13, 13, 12)$; details of implementation depend on the particular system. The boundary between these sublists is indicated by dotted lines at stage 1 in Fig. 6.4. After the first stage, the process is repeated on the sublists; but the selection of the median is based on the next significant bit, and so on. In this example, no exchanges take place during the third stage. The process is repeated until either all digits are processed, or sublists have only a single key.

It should be noted that the operations on the sublists are confined to areas defined by a continuously up-dated series of boundaries. This restriction may be removed if we make use of an observation that dates back to the days of mechanical card sorters; see, for example, Knuth (1973) (pp. 170–172) on sorting by distribution. The selections (compressions) are performed from the least significant bit, so that there is no need to keep the boundaries. The entire array is compressed at each stage. Figure

keys	binary	1st stage	2nd stage	3rd stage	4th stage	result
13	1101	0110	0100	0100	0100	4
7	0111	0100	1100	1100	0110	6
13	1101	1100	1101	1101	0111	7
6	0110	1101	1101	1101	1100	12
4	0100	0111	0110	0110	1101	13
12	1100	1101	0111	0111	1101	13
		↑	↑	↑	↑	

Fig. 6.5 Quicksort based on least significant bit

6.5 illustrates the sorting process based on the least significant bit for the above example. No exchanges take place during the second stage.

Clearly, in this approach the data are routed throughout the array at each stage, whereas the earlier approach involves shorter and shorter routeing operations.

In terms of computational complexity, the method has the same characteristics as the bit-level algorithms discussed in Chapter 2 (Section 2.7). The time is not dependent on the number of keys, as in word-based systems, but instead depends on the bit length of the largest key, and on the compression operations. The method is particularly effective when keys have short range, i.e. require a small number of binary digits. It is also easily adaptable to vector processors. For further discussion, the reader is referred to the paper by Boyle and Parkinson (1985), which also describes compression techniques employed on the DAP. The serial method related to the above techniques is referred to as the *radix exchange sort*; it uses two pointers—one at the front of the list and one at the back—and performs comparison exchanges on pairs of keys successively.

3.7 Address sort

Sorting by address can be faster than other serial methods, and can be carried out rapidly by some parallel switching networks. Given a list $L = (l_1, \ldots, l_n)$ of n elements (or keys), each element is transformed into a unique address and placed in an ordered output list corresponding to that address. After all the elements have been placed in the output list, the list is compressed to eliminate the empty locations, and sorting is thereby accomplished. The choice of a suitable sorting function is important to ensure both (i) that the elements are distributed evenly, with fewer empty locations remaining in the output list when sorting is complete, and (ii) that the transformation generally gives a unique address for each element. An efficient sorting function is clearly data-dependent, requiring prior knowledge of the size and number of elements. Despite this inconvenience, the address calculation is an effective sorting method and can break the $n \log_2 n$ barrier for the number of comparisons. The calculation of an address can be more complicated; however, with the use of switching networks it can be done in parallel and simultaneously routed to the destination.

For further reading see Isaac and Singleton (1956) and Feurzeig (1960).

3.8 Complete parallel sorting schemes

The complete parallel sorting schemes discussed below were arrived at when investigating the ordering of the eigenvalues resulting from various mobile schemes (described in Chapter 5, Section 5.7). Briefly, the latter involves a classification of the pairs (a, b) (with $1 \leq a < b \leq N$) into subsequences $L_1, L_2, \dots$ of roughly equal length. It was found that some such schemes sort the numbers $1, \dots, N$; in particular, one of them fully sorts every permutation of the numbers $1, \dots, N$ and can therefore be regarded as a sorting scheme in its own right. Here we provide a resumé of the paper by Davies and Modi (1986b) which discusses these issues in detail. We first introduce some terminology.

A list $L = ((a_1, b_1), \dots, (a_m, b_m))$ (without repetition) of some or all of the $\frac{1}{2}N(N-1)$ pairs (a, b) with $1 \leq a < b \leq N$ determines a (partially) sorting scheme as follows. Let N numbered locations be initially occupied by the integers $1, \dots, N$ in any order, the occupant of the nth location being $O_0(n)$. Perform successive 'exchanges', the mth being the interchange of the current occupant $O_{m-1}(a_m)$ of location a_m with $O_{m-1}(b_m)$ if the former is larger. We refer to L as *fully sorting* if there exists M such that $O_m(n) = n$ for $n = 1, \dots, N$ for every initial arrangement. The minimization of M for such schemes is discussed in the preceding sections. Here we consider schemes with

$$M = \tfrac{1}{2}N(N-1), \tag{8.1}$$

so that L contains all possible pairs; we call these *complete*. These schemes are certainly not competitive with those discussed previously. However, they may be of independent interest: one specific area of application is to consider whether a sorting of eigenvalues could be usefully and conveniently incorporated into the parallel Jacobi method (Chapter 5).

We start with an example with $N = 3$, and consider the six listings of (1 2), (1 3), (2 3):

(1 2)(2 3)(1 3) takes (2 3 1) and (3 2 1) to (2 1 3);

(2 3)(1 2)(1 3) takes (3 1 2) and (3 2 1) to (1 3 2).

In fact, of the six listings, only four are fully sorting. The proportion of complete schemes that are fully sorting decreases rapidly as N increases. There does not appear to be any simple test to ascertain whether or not a scheme is fully sorting. It is known (Knuth 1973, p. 224) that, if L sorts every sequence of 0's and 1's, then it is fully sorting. However, this merely reduces the number of initial arrangements to be tested from $N!$ to 2^N, and therefore does not help if N is large.

If a given complete scheme is applied repeatedly, then every initial arrangement is eventually sorted, because there will be an improvement at each application: a decrease in the number of misordered pairs. Here we consider schemes that sort in one application. The most obvious and the simplest of these is the lexicographical one:

$$((1, 2),\ldots, (1, N); (2, 3),\ldots, (2, N); \ldots; (N-1, N)) \tag{8.2}$$

which first brings 1 to the leading position, then 2 to the second position, and so on. The list (8.2) in reverse order is fully sorting. Unfortunately, these schemes are not efficient for parallel computation, as we indicate below.

Parallel sorting schemes

For implementation on a parallel computer, successive exchanges (a_i, b_i), (a_{i+1}, b_{i+1}), ... from the list L can be made simultaneously provided that none of these pairs have common elements. We thus have

$$L = (L_1, \ldots, L_K), \qquad L_k = ((a_1^k, b_1^k), \ldots, (a_{s(k)}^k, b_{s(k)}^k))$$
$$(k = 1, \ldots, K), \quad (8.3)$$

where each sublist L_k represents a set of exchanges that can be performed simultaneously. The $2s(k)$ elements comprising L_k in (8.3) are distinct, and consequently

$$2s(k) \leqslant \begin{cases} N & \text{if } N \text{ is even} \\ N-1 & \text{if } N \text{ is odd} \end{cases} \quad (k = 1, \ldots, K). \quad (8.4)$$

We are particularly interested in schemes for which K, the total number of parallel steps (*delay* in Knuth's terminology), is minimized.

By our completeness assumption (8.1)

$$M = \tfrac{1}{2} N(N-1) = s(1) + \ldots + s(K).$$

Hence, since the bounds in (8.4) are attainable for some lists L, it follows that the minimum value of K, written $K_{\min}$, is

$$K_{\min} = \begin{cases} N-1 & \text{if } N \text{ is even, with } s(k) = \tfrac{1}{2} N \text{ for all } k, \\ 0 & \text{if } N \text{ is odd, with } s(k) = \tfrac{1}{2}(N-1) \text{ for all } k. \end{cases}$$
$$(8.5)$$

(The minimum value of K is identical to the minimum number of parallel stages required in a single sweep of the parallel Jacobi method described in Chapter 5.)

The schemes discussed below also satisfy a further desideratum: the passage from L_k to L_{k+1} incurs minimum data movement, which is a particular advantage in parallel implementation. We describe three *alternative mobile* (AM) schemes which are formed as follows.

The numbers $1, \ldots, N$ are written down in a certain order: $x_1, \ldots, x_N$. Then two operations are applied alternately.

First operation Parentheses are inserted in one of the following three ways:

$$(x_1, x_2)(x_3, x_4) \ldots, \qquad x_1(x_2, x_3)(x_4, x_5) \ldots,$$
$$x_1)(x_2, x_3) \ldots (x_N, x_{N-1})(x_N,$$

the last being possible only if N is even, with (x_N, x_1) regarded as present. The bracketed pairs constitute the sublist L_1. (For the sake of convenience, we disregard the convention that pairs are written with the smaller element first.) Next, certain of these pairs are transposed and the parentheses dropped, giving a new sequence $(y_1, ..., y_N)$.

Second operation Parentheses are inserted in a different way, giving the second sublist L_2, and certain pairs are transposed, giving a third sequence $(z_1, ..., z_N)$.

The first operation is then applied to this, the second operation to the result, and so on. This often gives rise to a complete scheme.

(i) *Scheme introduced by Sameh (1971)*

This applies for even N, and can be exhibited as an AM scheme as follows. We take the case $N = 10$ for illustration.

$$L_1: \quad (8\ 1)(\mathbf{\textit{10\ 9}})(2\ 7)(4\ 5)(6\ 3)$$
$$L_2: \quad 1)(\mathbf{\textit{8\ 10}})(9\ 7)(2\ 5)(4\ 3)(6$$
$$L_3: \quad (6\ 8)(\mathbf{\textit{10\ 7}})(9\ 5)(2\ 3)(4\ 1)$$

etc.

Bold italicization indicates a pair not to be transposed. The scheme is easily seen to be complete, but is unfortunately not fully sorting for $N \geq 6$; for instance, with $N = 6$, it takes $(1\ 2\ 6\ 3\ 5\ 4)$ to $(1\ 2\ 3\ 4\ 6\ 5)$. It should be noted that the scheme was not designed for sorting but for describing the annihilation ordering of the pivots in the parallel Jacobi method, and the observation that it can be exhibited as an AM scheme is perhaps new. Further, we do not know whether the same operations applied to a different initial sequence might yield a fully sorting scheme. Such a variant is referred to as a *conjugate* of a given scheme, and may possess completely different sorting properties.

(ii) *Scheme introduced by Modi and Pryce (1985)*

This applies for every N; when N is even, there is one nonbracketed pair in each of $L_2, L_4, ...$, and therefore K is N rather than $N - 1$. For $N = 7$ and $N = 8$ respectively, this scheme, which is easily seen to be complete, is

L_1: (1 2)(3 4)(5 6)**7** L_1: (1 2)(3 4)(5 6)(7 8)
L_2: **2**(1 4)(3 6)(5 7) L_2: **2**(1 4)(3 6)(5 8)**7**
L_3: (2 4)(1 6)(3 7)**5** L_3: (2 4)(1 6)(3 8)(5 7)
etc.   etc.

The scheme is fully sorting. The proof is beyond the scope of this text, but is given by Davies and Modi (1986*b*).

(iii) *Scheme introduced by Davies and Modi (1986b)*

This applies for even N, say $N = 2P$, and can be described as follows. Let

$$L_1: (1\ 2)(3\ 4)(5\ 6)...,$$

then continually apply the cyclic permutation:

$$(2\quad 2P-1\quad 4\quad 2P-3\quad 6...3\quad 2P),$$

giving

L_2: $(1\quad 2P-1)(2P\quad 2P-3)(2P-2\quad 2P-5)\ldots$

L_3: $(1\ 4)(2\ 6)(3\ 8)...,$

and so on. This can be exhibited as an AM scheme as follows, taking $P = 4$ for illustration.

L_1: (**1 2**)(3 4)(5 6)(7 8)
L_2: **1**)(2 4)(3 6)(5 8)(**7**
L_3: (**1 4**)(2 6)(3 8)(6 7),
etc.

Viewed in this way, the scheme is a conjugate of one devised by Brent and Luk (1982), which begins with

$$L_1: (1\ 2)(5\ 6)(8\ 7)(4\ 3).$$

It is also a conjugate of Sameh's scheme, and can be derived from the Modi–Pryce scheme for $N - 1$. The scheme has been verified to be fully sorting for $P \le 7$, but the result has not been proved in general. Unfortunately, it does not have the property, possessed by the Modi–Pryce scheme, that inequalities $O_k(a) < O_k(b)$ once established are never subsequently lost.

Part 2

Numerical Linear Algebra

4 Solution of a system of linear algebraic equations

4.1 Introduction

This chapter covers two related topics: parallel techniques for solving linear equations and partial differential equations, both of which are of fundamental importance in research and industry.

To begin with, we examine in the abstract the solution of a system of linear equations

$$Ax = b, \tag{1.1}$$

where A is an $n \times n$ matrix and x and b are vectors of order n.

Direct techniques—the Gauss–Jordan method, WZ factorization, and QR factorization—which are particularly suited to parallel computation, are discussed in Section 4.2 without making any assumption as to the structure of A. Judgements based on serial techniques can often be misleading in a parallel environment. An inefficient or obsolete serial method may entail a large number of independent computations and so, potentially, a high degree of parallelism. The adaptation of serial techniques to a parallel environment is discussed with reference to the Gauss–Jordan method. We then proceed in Section 4.3 to describe two methods: the column-sweep algorithm and the recurrent-product algorithm for solving a triangular system of equations, which demonstrate the use of linear recurrences as well as serving to illustrate once again how parallelism can be extracted from serial methods.

The discussion of solution techniques is followed by a consideration of the way in which linear equations arise from partial differential equations. In the 1930s, these equations were solved without explicit reference to the matrix–vector form (1.1). This was because the kind of equations which resulted from finite-difference or finite-element discretization were sparse with band structure, and could quite profitably be expressed directly in terms of a two-dimensional grid of solutions (see, for example, Southwell's work on relaxation techniques, and various publications in the *American Journal of Applied Mathematics*). The subsequent emphasis on the matrix–vector form was in part due to the unprecedented availability of computers in the 1950s and the emergence of software packages requiring users to provide explicit descriptions of the resulting linear system. In engineering at least, the nature of the problem persuades one not to use the matrix–vector form as a straitjacket operation. In fact, solution techniques often reflect the nature of the problem, as, for example, in the Gauss–Seidel method, where overrelaxation has semiphysical meaning in the context of stress analysis. Thus the approach we have adopted is to start from partial differential equations and consider finite-difference and finite-element discretization of two-dimensional second-order elliptic equations.

The *iterative* techniques arising from finite-difference approximation are discussed in Section 4.4. In each case, we distinguish whether or not the approximate function values at grid points depend on the current iteration values at neighbouring points: (i) without interdependence, as in the point Jacobi method (not to be confused with the Jacobi method for the eigenvalue problem), (ii) *with* pointwise interdependence, as in the Gauss–Seidel method or successive-overrelaxation method, (iii) with pointwise interdependence, but where groups of points are updated, as in the line successive-overrelaxation and the alternating-direction implicit methods, which combine both direct and iterative techniques, and finally (iv) with pointwise interdependence, where groups of points are updated using explicit block-iterative schemes.

We then describe the finite-element process (Section 4.5) and discuss parallel techniques applicable to each phase of that process. The chapter concludes with a brief discussion of the conjugate-gradient method of solution, which can be readily adapted to parallel computation. Theoretically, the conjugate-gradient method requires a finite number of steps, and thus is most closely linked to direct solution techniques.

Naturally, the subject is too broad to be given detailed exposition in a short chapter such as this, and inevitably much has been left out. However, we do provide a resumé which gives some idea of the nature of research into the use of parallelism for a selected number of the algorithms mentioned above.

4.2 Factorization techniques

In this section we discuss three techniques, which are particularly suited to parallel computation, for solving linear algebraic systems. Before doing so, it should be noted that the choice of method depends on the type of parallel environment used. Most serial techniques (such as Gaussian elimination, or its variant LU decomposition) reduce the system of algebraic equations to a triangular form, and then use back substitution to solve the resulting system. On some parallel systems, however, this approach does not seem particularly promising. In the course of computation using Gaussian elimination, for example, the size of system under consideration is reduced at each stage, resulting in a considerable underutilization of processors. The Gauss–Jordan method, on the other hand, reduces the algebraic system straight to a diagonal form, and thus there is *no* back substitution phase. An interesting point here is the number of operations in each case: the Gauss–Jordan requires $\frac{1}{2}n^3 + O(n^2)$ multiplications while Gaussian elimination requires only $\frac{1}{3}n^3 + O(n^2)$. Thus, on a serial machine, where there is a direct relationship between the number of operations and run time, Gaussian elimination clearly wins. On a parallel machine, such as the DAP, however, serial operation count is not very useful, because in one parallel step (or termwise operation) several scalar operations can be performed simultaneously (see Chapter 1, Section 1.10). In terms of parallel operations, Gaussian elimination takes $O(n)$ termwise multiplications for the elimination and a further $O(n)$ termwise multiplications for the back-substitution phase, whereas Gauss–Jordan takes only $O(n)$ termwise multiplications in all. Thus the Gauss–Jordan method provides the solution in the same number of operations as the elimination phase of the Gaussian elimination method. Further, the above argument would seem to imply that triangular forms are essentially a byproduct of the way we have conventionally solved the problem, and their use in a parallel environment, with sufficient processors available, is not easily justifiable. A number of authors have nevertheless devised for triangular systems some neat and clever solution techniques which also demonstrate the use of linear recurrences: two of these, the *column-sweep* algorithm and the *recurrent-product* algorithm, we will discuss in Section 4.3. In the first instance, however, we present the following three methods for solving linear algebraic systems, without making any assumptions as to the structure of the matrix: (a) the Gauss–Jordan method, (b) WZ factorization, and (c) QR factorization.

(a) *Gauss–Jordan method*

The Gauss–Jordan method, a slight variation on the schoolboy approach to solving simultaneous equations, may be illustrated by way of a simple example. We present the method in its entirety, although it will be obvious that a number of short cuts could be made.

Suppose we wish to solve the following 4×4 set of linear equations for x_1, x_2, x_3, and x_4.

$$2x_1 + 2x_2 + 4x_3 + 2x_4 = 76 \qquad (1,1)$$

$$x_1 + 3x_2 + 2x_3 + x_4 = 52 \qquad (1,2)$$

$$3x_1 + 2x_2 + 2x_3 + x_4 = 57 \qquad (1,3)$$

$$x_1 + x_2 + 3x_3 + 3x_4 = 64. \qquad (1,4)$$

Divide $(1,1)$ by the coefficient of x_1 to give $(2,1)$, and then perform the operations

$$(2,2) \leftarrow (1,2) - (2,1),$$

$$(2,3) \leftarrow (1,3) - 3(2,1),$$

$$(2,4) \leftarrow (1,4) - (2,1),$$

to give

$$x_1 + x_2 + 2x_3 + x_4 = 38 \qquad (2,1)$$

$$2x_2 + 0x_3 + 0x_4 = 14 \qquad (2,2)$$

$$-x_2 - 4x_3 - 2x_4 = -57 \qquad (2,3)$$

$$0x_2 + x_3 + 2x_4 = 26. \qquad (2,4)$$

Next divide $(2,2)$ by the coefficient of x_2 to give $(3,2)$, and perform the operation

$$(3,3) \leftarrow (2,3) + (3,2)$$

(the coefficient of x_2 in (2,4) is already zero), to give

$$x_1 + x_2 + 2x_3 + x_4 = 38 \qquad (3,1)$$

$$x_2 + 0x_3 + 0x_4 = 7 \qquad (3,2)$$

$$-4x_3 - 2x_4 = -50 \qquad (3,3)$$

$$x_3 + 2x_4 = 26. \qquad (3,4)$$

The sequence of operations so far comprises typical Gaussian-elimination steps, where the variables are eliminated below the pivot row. In the Gauss–Jordan method, we annihilate the elements above the diagonal as well as below. Thus, in the previous sequence of steps, we would also perform the step

$$(4,1) \leftarrow (3,1) - (3,2),$$

which annihilates the coefficient of x_2 in (3,1), to give

$$x_1 \quad + 2x_3 + x_4 = 31 \qquad (4,1)$$

$$x_2 + 0x_3 + 0x_4 = 7 \qquad (4,2)$$

$$-4x_3 - 2x_4 = -50 \qquad (4,3)$$

$$x_3 + 2x_4 = 26. \qquad (4,4)$$

Next divide (4,3) by the coefficient of x_3 to give the pivot row (5,3), and perform the following operations (i.e. add/subtract multiples of the pivot row from *all* other rows):

$$(5,1) \leftarrow (4,1) - 2(5,3)$$

$$(5,4) \leftarrow (4,4) - (5,3)$$

(the coefficient of x_3 in (4,2) is already zero), to give

$$x_1 \qquad + 0x_4 = 6 \qquad (5,1)$$

$$x_2 \quad + 0x_4 = 7 \qquad (5,2)$$

$$x_3 + \tfrac{1}{2}x_4 = \tfrac{25}{2} \qquad (5,3)$$

$$\tfrac{3}{2}x_4 = \tfrac{27}{2}. \qquad (5,4)$$

Finally, divide (5,4) by the coefficient of x_4 to give (6,4), and then perform the operation

$$(6,3) \leftarrow (5,3) - \tfrac{1}{2}(5,4)$$

(the coefficients of x_4 in (5,1) and (5,2) are already zero), to give

$$x_1 \qquad\qquad\qquad = 6 \qquad\qquad (6,1)$$

$$x_2 \qquad\qquad = 7 \qquad\qquad (6,2)$$

$$x_3 \quad = 8 \qquad\qquad (6,3)$$

$$x_4 = 9. \qquad\qquad (6,4)$$

Back substitution is unnecessary, and this is the major advantage over Gaussian elimination. If we define a *step* to consist of the simultaneous annihilation of a specified coefficient from every row other than the pivot row, then this requires $O(1)$ termwise multiplications on a parallel system. Thus the entire process is completed in $O(n)$ termwise multiplications. The weakness of the method is that it may sometimes prove numerically unstable, and some form of pivoting (such as *column pivoting*) should therefore be incorporated. The extension of this method to a banded set of linear algebraic equations has been carried out by Biswas and Amaratunga (1987), who demonstrate its effectiveness on an SIMD machine.

(b) *WZ factorization*

WZ factorization is an interesting variant of the elimination method. The idea is to factorize A into the form WZ, as defined below, where the associated sequence of computational operations naturally appeals to parallelism. The method is asserted to have the same numerical stability as Gaussian elimination when the matrix is symmetric positive definite, or diagonally dominant. A brief description of the method is presented below.

We solve the linear system of equations $Ax = b$ by factorizing A in the form

$$A = WZ;$$

here W and Z have the forms

$$
W = \begin{bmatrix}
1 & & & & & 0 \\
w_{2,1} & 1 & & 0 & 0 & w_{2,n} \\
\vdots & & 1 & & & \vdots \\
w_{n-1,1} & 0 & & 0 & 1 & w_{n-1,n} \\
0 & & & & & 1
\end{bmatrix},
$$

$$Z = \begin{bmatrix} z_{1,1} & & & \cdots & & z_{1,n} \\ 0 & z_{2,2} & & \cdots & z_{2,n-1} & 0 \\ \vdots & 0 & \ddots & z_{i,i} & 0 & \vdots \\ 0 & & \ddots & & \ddots & 0 \\ z_{n,1} & & & \cdots & & z_{n,n} \end{bmatrix},$$

where

$$W_{i,j} = \begin{cases} 1 & \text{if } i = j, \\ 0 & \text{if } i+1 \leqslant j \leqslant n-i+1 \text{ or } n-i+1 \leqslant j \leqslant i-1, \end{cases}$$

$$Z_{i,j} = 0 \quad \text{if } j+1 \leqslant i \leqslant n-j \text{ or } n-j+2 \leqslant i \leqslant j-1.$$

The other elements of W and Z are computed as follows. Initially:

(a)   the first and last rows of Z are given by

$$z_{1,j} = a_{1,j} \text{ and } z_{n,j} = a_{n,j} \quad (j = 1, \ldots, n);$$

(b)   for the outermost columns of W, we solve $n - 2$ sets of 2×2 linear systems

$$z_{1,1}w_{i,1} + z_{n,1}w_{i,n} = a_{i,1},$$

$$z_{1,n}w_{i,1} + z_{n,n}w_{i,n} = a_{i,n},$$

for $w_{i,1}$ and $w_{i,n}$ $(i = 2, \ldots, n-1)$.

Generally, by recursion on $k = 1, \ldots, \lfloor \frac{1}{2}(n+1) \rfloor$, the elements $w_{i,v}$ and $z_{v,j}$ $(v = k, n-k+1)$ are found by setting

$$z_{v,j} = a_{v,j} - \left(\sum_{l=1}^{k-1} + \sum_{l=n-k+2}^{n} \right) w_{v,l} z_{l,j}$$

$$(v = k, n-k+1; j = k, \ldots, n-k+1),$$

and, for $i = k+1, \ldots, n-k$, solving for $w_{i,v}$ the equation pair

$$z_{k,v}w_{i,k} + z_{n-k+1,v}w_{i,n-k+1} = a_{i,v} - \left(\sum_{l=1}^{k-1} + \sum_{l=n-k+2}^{n} \right) w_{i,l} z_{l,v}$$

$$(v = k, n-k+1).$$

At least $O(n)$ parallel steps are required to compute the elements of W and Z.

We note that, under certain circumstances, the WZ factorization may give spurious results. Consider, for example, the nonsingular matrix

$$A = \begin{bmatrix} 1 & 1 & 1 & 1 \\ 0 & 0 & 0 & 1 \\ 0 & 0 & 1 & 1 \\ 1 & 0 & 0 & 1 \end{bmatrix},$$

and suppose that we seek to factorize A as $A = WZ$, where

$$W = \begin{bmatrix} 1 & 0 & 0 & 0 \\ w_{21} & 1 & 0 & w_{24} \\ w_{31} & 0 & 1 & w_{34} \\ 0 & 0 & 0 & 1 \end{bmatrix}, \quad Z = \begin{bmatrix} z_{11} & z_{12} & z_{13} & z_{14} \\ 0 & z_{22} & z_{23} & 0 \\ 0 & z_{32} & z_{33} & 0 \\ z_{41} & z_{42} & z_{43} & z_{44} \end{bmatrix}.$$

We see directly that $z_{11} = z_{14} = z_{41} = z_{44} = 1$. A comparison of the $(2,1)$ and $(2,4)$ entries of A and WZ reveals the contradiction

$$\left. \begin{matrix} w_{21} + w_{24} = 0 \\ w_{21} + w_{24} = 1 \end{matrix} \right\}.$$

Thus, if $\bar{A}$ is a matrix that differs from A by a matrix whose entries are comparable in size with the machine tolerance, then $\bar{A}$ may have well-defined W and Z factors that are (just) computable. While the basic WZ algorithm would not crash, the output would be unreliable, even though $\bar{A}$ is well conditioned and not nearly singular. We therefore need to incorporate into the basic algorithm, at each stage k, a check on the size of

$$\delta_k = \det \begin{bmatrix} z_{k,k} & z_{k,n-k+1} \\ z_{n-k+1,k} & z_{n-k+1,n-k+1} \end{bmatrix}$$

to ensure that the algorithm is rejected if $|\delta_k|$ is too small.

Having computed the W and Z factors, we then proceed to solve (1.1) as follows. Rewrite $Ax = b$ as $WZx = b$, so that we first compute an n-element vector p:

$$Wp = b; \tag{2.1}$$

then we solve for x via

$$Zx = p. \qquad (2.2)$$

From (2.1), we have

$$\begin{bmatrix} 1 & & & & 0 \\ w_{2,1} & \cdot & & \cdot & w_{2,n} \\ \vdots & & 1 & & \vdots \\ w_{n-1,1} & \cdot & & \cdot & w_{n-1,2} \\ 0 & & & & 1 \end{bmatrix} \begin{bmatrix} p_1 \\ p_2 \\ \vdots \\ p_{n-1} \\ p_n \end{bmatrix} = \begin{bmatrix} b_1 \\ b_2 \\ \vdots \\ b_{n-1} \\ b_n \end{bmatrix},$$

from which we compute p_1 and p_n first, then p_2 and p_{n-1}, and so on—each time working from the top and the bottom. At the ith step, we have

$$p_i = b_i^{(i)}, \qquad p_{n-i+1} = b_{n-i+2}^{(i)},$$

where $b_j^{(1)} = b_j$ and

$$b_j^{(i+1)} = b_j^{(i)} - w_{j,i}p_i - w_{j,n-i+1}p_{n-i+1} \quad (j = i+1, \ldots, n-i).$$

The computation of x in (2.2) is similar (for evaluating recurrence relations in parallel, see Chapter 2, Sections 2.5 and 2.6). Further details of the method, and its extension to banded linear systems, are discussed by Evans *et al.* (1981); its implementation on the DAP is described by Hellier (1982).

Iterative methods similar to the point Jacobi and successive-overrelaxation (Section 4.4) may be devised by quadrant–diagonal partitioning (QDP): $A = X - W' - Z'$, where X has nonzeros only on the main and cross diagonals, and $-W'$ and $-Z'$ have zeros on the main and cross diagonals, and nonzeros only in the shaded regions indicated in Fig. 2.1. The elements of X, $-W'$, and $-Z'$ are given by:

$$(X)_{ij} = \begin{cases} 0 & \text{if } i \neq j \text{ and } i \neq n - j + 1, \\ a_{ij} & \text{otherwise,} \end{cases}$$

$$(-W')_{ij} = \begin{cases} 0 & \text{if } i \leq j \leq n - i + 1 \text{ or } n - i + 1 \leq j \leq i, \\ a_{ij} & \text{otherwise,} \end{cases}$$

$$(-Z')_{ij} = \begin{cases} 0 & \text{if } j \leq i \leq n - j + 1 \text{ or } n - j + 1 \leq i \leq j, \\ a_{ij} & \text{otherwise.} \end{cases}$$

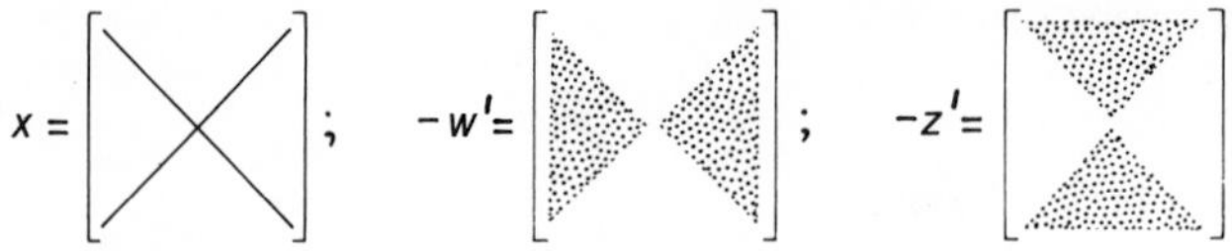

Fig. 2.1 Quadrant–diagonal partitioning of A

The schemes of Section 4.4 are based on partitioning A into D, L, and U, where L and U are strictly lower triangular and strictly upper triangular matrices respectively, and D is a diagonal matrix. Similarly, we may devise the following schemes by applying the quadrant–diagonal partitioning $A = X - W' - Z'$ to the linear system $Ax = b$.

Jacobi QDP method:

$$Xx^{(p+1)} = (W' + Z')x^{(p)} + b, \qquad (2.3)$$

Successive QDP method:

$$(X - W')x^{(p+1)} = Z'x^{(p)} + b, \qquad (2.4)$$

Successive-overrelaxation QDP method:

$$(X - \omega W')x^{(p+1)} = [\omega Z' + (1 - \omega)X]x^{(p)} + \omega b, \qquad (2.5)$$

where ω is the relaxation factor.

The schemes described by (2.3)–(2.5) can be regarded as (2×2) block-iterative methods, and lead naturally to parallelism. A number of useful convergence properties may be proved: (i) the Jacobi method converges if A is irreducible and diagonally dominant, (ii) the successive QDP method converges if A is irreducible and possesses weak diagonal dominance, or if A is real and positive definite, and finally (iii) the successive-overrelaxation QDP method converges for $0 < \omega < 2$ and corresponds to a (2×2) block-successive-overrelaxation method with

$$\omega = \frac{2}{1 + [1 - \rho^2(J)]^{\frac{1}{2}}},$$

where $\rho(J)$ is the spectral radius of the iteration matrix in the corresponding block Jacobi method. (For an $n \times n$ complex matrix A with eigenvalues λ_i $(1 \le i \le n)$, the spectral radius of A is defined as $\rho(A) = \max\{|\lambda_i| : 1 \le i \le n\}$.) Further details may be found in Evans (1982).

(c) *QR factorization*

QR factorization is discussed in detail in Chapter 6. Here we consider its application in the solution of linear algebraic equations. Strategies similar to those discussed below can be applied to perform low-rank update, which is particularly useful in optimization and solving field-analysis problems, where small changes have been made to the discretized region under consideration.

An $n \times n$ matrix A is factorized as

$$A = QR,$$

where Q is an $n \times n$ orthogonal matrix and R is an $n \times n$ upper triangular matrix, namely

$$R = \begin{bmatrix} r_{11} & r_{12} & \cdots & r_{1n} \\ & r_{22} & \cdots & r_{2n} \\ & & \ddots & \vdots \\ \mathbf{0} & & & r_{nn} \end{bmatrix}.$$

(This definition is a special case of QR factorization for a square matrix A; the usual definition for an $m \times n$ matrix is given in Chapter 6.)

We can use Givens' plane rotations or Householder's reflections to determine Q and R. The former method determines R by calculating $Q^{\mathrm{T}}A$, where Q is the product of a number of plane rotations, each of which annihilates an element of A below the main diagonal without destroying the previously introduced zeros. In order to annihilate the (i, j)th element of A, we premultiply the ith and $(i-1)$th rows by $p = \begin{bmatrix} c & s \\ -s & c \end{bmatrix}$, where $c = \cos\theta$ and $s = \sin\theta$, the angles of rotation being determined by using the formulae

$$s = a_{i,j}/(a_{i,j}^2 + a_{i-1,j}^2)^{\frac{1}{2}}, \qquad c = a_{i-1,j}/(a_{i,j}^2 + a_{i-1,j}^2)^{\frac{1}{2}}.$$

In a parallel method, several rotations are applied simultaneously, as illustrated in Fig. 2.2 for $n = 8$. An integer r is

```
X
3  X
2  5  X
2  4  7  X
1  3  6  8  X
1  3  5  7  9  X
1  2  4  6  8  10  X
1  2  3  5  7  9  11  X
```

Fig. 2.2 Greedy algorithm applied to an 8×8 matrix

entered where zeros are created at the rth step. The ordering of annihilation is that of the 'greedy' algorithm† described in Section 6.3 of Chapter 6. As seen from Fig. 2.2, the scheme takes 11 steps to accomplish the factorization. We could equally have chosen Sameh and Gentleman's modification of the serial ordering, taking 13 steps. The choice in practice is largely governed by communications overheads, which can vary significantly from one system to another. The routeing network connecting the processing elements, as well as the availability of reconfigurable topology and the granularity of the processing, strongly influences the choice of ordering.

We can therefore proceed to solve for x in (1.1) as follows. Rewrite $Ax = b$ as $QRx = b$, so that

$$Rx = Q^Tb.$$

There is no need to compute Q explicitly, simply the product Q^Tb. The elements of x can be obtained by solving the triangular system by back substitution (Section 4.3).

4.3 Triangular systems of linear equations

In this section, we describe the column-sweep and the recurrent-product algorithm for solving a triangular system of linear equations; such a system often arises in the final phase of solving dense systems, as in the case of the QR factorization method discussed in the previous section. The methods both demonstrate the use of linear recurrence and illustrate how parallelism can be

---

† The author first termed this the 'new scheme', as in Chapter 6 and in Modi (1982). With some justification perhaps, Cosnard and Robert (1986) subsequently refer to it as the 'greedy' algorithm.

exploited in a serial method. The column-sweep algorithm requires $O(n)$ steps to solve an $n \times n$ triangular system using $O(n)$ processors, whereas the recurrent-product algorithm requires $O((\log_2 n)^2)$ steps using $O(n^3)$ processors—which may not be possible for a realistic value of n ($n > 100$) on some currently available systems.

We begin by defining a linear recurrence system $R\langle n, m \rangle$ of order m for n equations by

$$R\langle n,m \rangle : \quad x_k = \begin{cases} 0 & \text{if } k \leq 0, \\ b_k + \displaystyle\sum_{j=k-m}^{k-1} a_{kj} x_j & \text{if } 1 \leq k \leq n, \end{cases} \quad (3.1)$$

where $m \leq n - 1$. If we write $A = [a_{ik}]$, where $a_{ik} = 0$ for $i \leq k$ or $i - k > m$, and $x = [x_1, \ldots, x_n]^\mathsf{T}$ and $b = [b_1, \ldots, b_n]^\mathsf{T}$, then (3.1) can be written as

$$x = Ax + b, \quad (3.2)$$

where A is strictly lower triangular. Thus $R\langle 4,3 \rangle$, for example, has the form

$$x_1 = b_1$$
$$x_2 = b_2 + a_{21}x_1$$
$$x_3 = b_3 + a_{31}x_1 + a_{32}x_2$$
$$x_4 = b_4 + a_{41}x_1 + a_{42}x_2 + a_{43}x_3.$$

For $m < n - 1$, we obtain a banded system.

We can express a nondegenerate triangular system of equations $\tilde{A}x = \tilde{b}$ in the form (3.2) as follows. Rewrite $\tilde{A}x = \tilde{b}$ as $(I - A)x = b$, where $(A)_{ij} = \delta_{ij} - \tilde{a}_{ij}/\tilde{a}_{ii}$ and $(b)_i = \tilde{b}_i/\tilde{a}_{ii}$ ($i,j = 1, \ldots, n$). Then we obtain

$$x = Ax + b. \quad (3.3)$$

In order to illustrate the *column-sweep* algorithm, we take a 4×4 system: $\tilde{A}x = \tilde{b}$ with

$$\tilde{A} = \begin{bmatrix} 1 & & & \\ -a_{21} & 1 & & \\ -a_{31} & -a_{32} & 1 & \\ -a_{41} & -a_{42} & -a_{43} & 1 \end{bmatrix}, \quad x = \begin{bmatrix} x_1 \\ x_2 \\ x_3 \\ x_4 \end{bmatrix}, \quad \tilde{b} = \begin{bmatrix} \tilde{b}_1 \\ \tilde{b}_2 \\ \tilde{b}_3 \\ \tilde{b}_4 \end{bmatrix}.$$

Using (3.3), we obtain the set of equations

$$
\begin{aligned}
x_1 &= b_1 \\
x_2 &= \boxed{b_2 + a_{21}x_1} \\
x_3 &= \boxed{b_3 + a_{31}x_1} + a_{32}x_2 \\
x_4 &= \boxed{b_4 + a_{41}x_1} + a_{42}x_2 + a_{43}x_3,
\end{aligned}
\qquad (3.4)
$$

from which we can evaluate x_1, x_2, x_3, and x_4 in parallel as follows.

Step 1 Evaluate, in parallel, expressions of the form $b_i^{(1)} = b_i + a_{i1}x_1$, for $i = 2, \ldots, n$, where $x_1 = b_1$ is known—the computation is indicated by dotted lines in eqn (3.4). There now remain $n - 2$ equations to solve.

Step 2 Evaluate, in parallel, expressions of the form $b_i^{(2)} = b_i^{(1)} + a_{i2}x_2$, for $i = 3, \ldots, n$, where x_1 and x_2 are known; the computation is indicated by solid lines in eqn (3.4). There now remain $n - 3$ equations to solve.

. . .

Step k Evaluate, in parallel, expressions of the form $b_i^{(k)} = b_i^{(k-1)} + a_{ik}x_k$, for $i = k + 1, \ldots, n$, where $x_1, \ldots, x_k$ are known.

Clearly, for an $n \times n$ system, the method requires $n - 1$ steps and $n - 1$ processors at the first step, and fewer thereafter. It is particularly useful when relatively few processors ($\leq n$) are available. For a system with a relatively large number of processors—of order n^3—we can express the solution x as a product of elementary matrices that can easily be computed using the recursive-doubling technique described in Section 2.5 of Chapter 2. This method, which is referred to as the *recurrent-product* algorithm, is due to Sameh and Brent (1977) and proceeds by writing (3.2) as

$$ x = (I - A)^{-1}b; \qquad (3.5) $$

here the $n \times n$ matrix A is strictly lower triangular and $(I - A)^{-1}$ can be expressed in the product form (Householder 1974)

$$ (I - A)^{-1} = \prod_{i=1}^{n-1} M_{n-i}, $$

where

$$M_i = \begin{bmatrix} 1 & & & & & & \\ & \ddots & & & & 0 & \\ & & 1 & & & & \\ & & a_{i+1,i} & 1 & & & \\ & 0 & \vdots & & \ddots & & \\ & & a_{n,i} & 0 & & 1 \end{bmatrix}.$$

The evaluation of the matrix product $\prod_{i=1}^{n-1} M_{n-i}$ can be carried out in $O((\log_2 n)^2)$ steps, using the recursive-doubling technique (see the theorem below).

Example Suppose we take a banded lower triangular system $R\langle 5, 2 \rangle$; then, with the help of (3.2), we obtain $x = Ax + b$, where

$$x_1 = b_1$$
$$x_2 = b_2 + a_{21}x_1$$
$$x_3 = b_3 + a_{31}x_1 + a_{32}x_2$$
$$x_4 = b_4 \qquad\qquad + a_{42}x_2 + a_{43}x_3$$
$$x_5 = b_5 \qquad\qquad\qquad\qquad + a_{53}x_3 + a_{54}x_4.$$

Hence

$$x = M_4 M_3 M_2 M_1 b, \tag{3.6}$$

where

$$M_4 = \begin{bmatrix} 1 & & & & \\ & 1 & & 0 & \\ & & 1 & & \\ & 0 & & 1 & \\ & & & a_{54} & 1 \end{bmatrix}, \quad M_3 = \begin{bmatrix} 1 & & & & \\ & 1 & & 0 & \\ & & 1 & & \\ & 0 & a_{43} & 1 & \\ & & a_{53} & 0 & 1 \end{bmatrix},$$

$$M_2 = \begin{bmatrix} 1 & & & & \\ 0 & 1 & & 0 & \\ 0 & a_{32} & 1 & & \\ 0 & a_{42} & 0 & 1 & \\ 0 & 0 & 0 & 0 & 1 \end{bmatrix}, \quad M_1 = \begin{bmatrix} 1 & & & & \\ a_{21} & 1 & & 0 & \\ a_{31} & 0 & 1 & & \\ & 0 & & 1 & \\ & & & & 1 \end{bmatrix}.$$

Using recursive doubling, the matrix product in (3.6) can be evaluated as

$$(((M_4 M_3)(M_2 M_1))b),$$

where the rule is that we compute all the inner brackets at step 1, do all 'next' brackets at Step 2, and so on. Thus, in this case, three parallel steps are required.

An upper bound in terms of a number p of processors and the number T_p of steps is indicated by the following.

Theorem (Kuck 1978) A linear recurrence system $R\langle n, m\rangle$ can be evaluated on p processors in T_p steps, where $T_p < (2 + \log_2 m)\log_2 n - \frac{1}{2}(1 + \log_2 m)\log_2 m$ and

$$p \leq \begin{cases} \frac{1}{2}m(m+1)n + O(m^3) & \text{for } 1 \leq m \leq \frac{1}{2}n, \\ n^3/68 \quad\quad\; + O(n^2) & \text{for } \frac{1}{2}n \leq m \leq n - 1. \end{cases} \qquad \Box$$

For the case $R\langle n, 2\rangle$, with large n, we have $T_p \leq 3\log_2 n - 1$ and $p \leq 3n - 8$. Thus, for the above example $R\langle 5, 2\rangle$, we get $T_p \leq 6$ and $p \leq 7$.

Numerically, the *recurrent-product* algorithm is less stable than the corresponding sequential method. An error analysis by Sameh and Brent (1977) shows that, if the computed solution is $\bar{x}$, then $(A + \delta A)\bar{x} = b$, where

$$\|\delta A\| \leq O(n^2 \log_2 n)K^2(A)\|A\|\varepsilon,$$

$K(A)$ is the condition number of A, $\|\bullet\|$ denotes the ∞-norm, and ε is the unit roundoff. For the sequential algorithm, Wilkinson (1963) has shown that

$$\|\delta A\| \leq n \|A\|\varepsilon.$$

Experimental evidence shows that, usually, identical answers are obtained in each case. In general, however, the parallel algorithm is less stable. On the other hand, the *column-sweep* algorithm, which is much slower, has the same error bound on δA as the corresponding sequential method, and it may be preferable for certain ill-conditioned systems. A cautionary note for users intending to program the recurrent-product algorithm is to be aware of the communication costs involved. Unless the interprocessor connectivity matches the communication incurred in the method reasonably well, the overheads may represent a significant proportion of the overall cost.

A number of other algorithms share this characteristic with the recurrent-product algorithm, taking $O((\log_2 n)^2)$ steps with $O(n^3)$ processors. Borodin and Munro (1975) propose a method that uses partitioning to invert a lower triangular matrix A as follows. Partition $A = \begin{bmatrix} A_1 & 0 \\ A_2 & A_3 \end{bmatrix}$, where A_1 and A_3 are lower triangular matrices, and A_2 is full, so that

$$A^{-1} = \begin{bmatrix} A_1^{-1} & 0 \\ -A_3^{-1} A_2 A_1^{-1} & A_3^{-1} \end{bmatrix}.$$

The method consists of two stages: (i) simultaneously invert A_1 and A_3 and solve $A_3 Y = A_2$ for Y, and (ii) multiply Y and A_1^{-1}. Such computations would clearly suit a small coarse-grain MIMD system where individual processors can be allocated the separate tasks of inversion, multiplication, etc. Methods which can be implemented on fewer than $O(n^3)$ processors are discussed by Hyafil and Kung (1977) and Chen (1975). For the case of Toeplitz triangular systems, the number of processors can be reduced to $O(n^2)$.

4.4 Finite-difference solution of partial differential equations

Finite-difference approximations form the basis of many iterative methods commonly used in solving partial differential equations. Their application ranges from the scientific to the commercial, from weather forecasting, simulation, and modelling aircraft design, to econometrics, and they all require vast amounts of computer time. It is pertinent therefore to consider their evaluation on a parallel computer. A procedure is defined here as iterative when we first estimate the value of some function ϕ at all mesh points, and then use a difference equation as the basis for calculating an improved value. After outlining the general principle, we will discuss some well-known techniques: point Jacobi, Gauss–Seidel, successive-overrelaxation (SOR), line successive-overrelaxation (LSOR), and block-iterative schemes. These schemes are inherently parallel, and the problem is simply to organize them conveniently on an array of processors.

The special case of the two-dimensional second-order partial differential equation for $\phi(x, y)$ may be written as

$$a\frac{\partial^2\phi}{\partial x^2} + b\frac{\partial^2\phi}{\partial x\,\partial y} + c\frac{\partial^2\phi}{\partial y^2} + d\frac{\partial\phi}{\partial x} + e\frac{\partial\theta}{\partial y} + f\phi + g = 0,$$

where $a, \ldots, g$ are continuous real functions of the independent variables x and y, and dependent variable ϕ. Three types of partial differential equation are distinguished: *elliptic*, for which $b^2 - 4ac < 0$, *parabolic*, for which $b^2 - 4ac = 0$, and *hyperbolic*, for which $b^2 - 4ac > 0$. In the following discussion we shall consider elliptic equations in two dimensions, but the techniques may be extended to higher dimensions.

The region R of integration of the elliptic equation is bounded by a closed curve C, along which there may be defined boundary conditions on $\phi(x, y)$ or its derivatives (Fig. 4.1). In the application of finite-difference methods, the region is discretized into small rectangular regions formed by sets of equally spaced lines parallel to the x and y axes. An approximate solution is found at the intersection of these lines, i.e. at the *grid points*, by approximating spatial derivatives of ϕ at grid point (i, j) by linear functions of the values of ϕ at grid point (i, j) and at certain neighbouring grid points (indicated by circles in Fig. 4.1). The application of finite differences at each grid point thus gives an algebraic equation. If there are p^2 grid points, then the resulting linear system consists of p^2 equations and p^2 unknowns. The solutions provide approximate values for ϕ at the specified grid

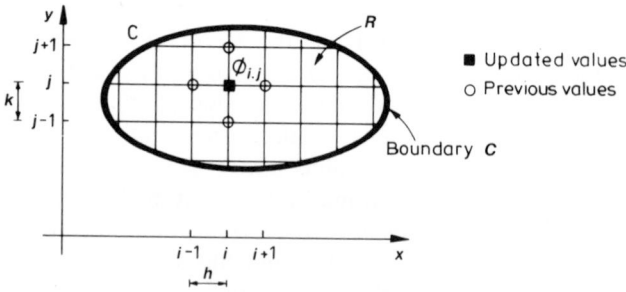

Fig. 4.1 Finite-difference discretization

points, and interpolation techniques can be used to obtain solutions at all points in R.

It is emphasized that the approximate method expounded here may yield accurate results. The derivatives at any given point are approximated by difference quotients *over intervals given by the grid spacing*; thus $\partial \phi / \partial x$ is replaced by $\delta \phi / \delta x$, where δx —effectively the grid spacing—can be made small enough in principle to achieve any desired accuracy. In practice, increased execution time and numerical roundoff error may be important considerations as the grid spacing is refined. For further discussion, the reader is referred to Smith (1978).

We next derive the finite-difference approximation to partial derivatives of $u(x, y)$, apply these results to approximate the Poisson equation

$$\frac{\partial^2 u}{\partial x^2} + \frac{\partial^2 u}{\partial y^2} = q, \tag{4.1}$$

with boundary conditions $u(x, y) = u_0(x, y)$ on C, and finally consider the evaluation using an array of processors. In engineering applications, (4.1) describes steady-state and diffusion phenomena. For instance, $u(x, y)$ might denote the steady-state temperature distribution in a two-dimensional isotropic homogeneous body in some closed region R in the (x, y) plane, and $q(x, y)$ would then describe the heat sources or sinks throughout that body.

Using Taylor's theorem in two dimensions, we have

$$u(x + y, y + k)$$
$$= u(x, y) + \frac{\partial u}{\partial x} h + \frac{\partial u}{\partial y} k + \tfrac{1}{2}\left(\frac{\partial^2 u}{\partial x^2} h^2 + 2 \frac{\partial^2 u}{\partial x \, \partial y} kh + \frac{\partial^2 u}{\partial y^2} k^2\right) + \cdots,$$
$$\tag{4.2}$$

and, if we keep y constant, then (4.2) reduces to

$$u(x + h, y) = u(x, y) + \frac{\partial u}{\partial x} h + \frac{1}{2} \frac{\partial^2 u}{\partial x^2} h^2 + \cdots, \tag{4.3}$$

$$u(x - h, y) = u(x, y) - \frac{\partial u}{\partial x} h + \frac{1}{2} \frac{\partial^2 u}{\partial x^2} h^2 - \cdots. \tag{4.4}$$

Adding (4.3) and (4.4), and ignoring $O(h^4)$ terms, gives

$$\frac{\partial^2 u}{\partial x^2} = \frac{u(x+h, y) - 2u(x, y) + u(x-h, y)}{h^2}. \qquad (4.5)$$

We then subdivide the region in the (x, y) plane into a number of equal rectangles of dimensions $\delta x = h$ and $\delta y = k$, as shown in Fig. 4.1, and represent the (x, y) coordinates by $x = ih$ and $y = jk$. This gives

$$\frac{\partial^2 u}{\partial x^2} = \frac{u((i+1)h, jk) - 2u(ih, jk) + u((i-1)h, jk)}{h^2},$$

which is commonly written as

$$\frac{\partial^2 u}{\partial x^2} = \frac{u_{i+1,j} - 2u_{i,j} + u_{i-1,j}}{h^2}. \qquad (4.6)$$

Similarly, if we keep x constant in (4.2), we obtain

$$\frac{\partial^2 u}{\partial y^2} = \frac{u_{i,j+1} - 2u_{i,j} + u_{i,j-1}}{k^2}. \qquad (4.7)$$

The division of the region described by (4.1) into unit squares (i.e. $k = h = 1$) gives, using (4.6) and (4.7), the finite-difference approximation to the Poisson equation:

$$u_{i,j} = \tfrac{1}{4}(u_{i-1,j} + u_{i+1,j} + u_{i,j-1} + u_{i,j+1} - q_{ij}). \qquad (4.8)$$

We now discuss three iterative schemes for evaluating (4.8) in parallel using an SIMD model of computation: i.e. identical instructions are applied to a set of data, with the processors interconnected to form a lattice where each processor is connected to four adjacent neighbours (Chapter 1, Section 1.7).

(i)   Point Jacobi method

The word 'point' indicates that the update is point-by-point (as opposed to line-by-line or block update); see (iv) and (v) below. The iterative procedure (first considered by Jacobi in 1845) expresses the next iterative values for each grid point in terms of known iterative values at adjacent points. The scheme is defined by

$$u_{i,j}^{(p+1)} = \tfrac{1}{4}(u_{i-1,j}^{(p)} + u_{i+1,j}^{(p)} + u_{i,j-1}^{(p)} + u_{i,j+1}^{(p)} - q_{i,j}). \qquad (4.9)$$

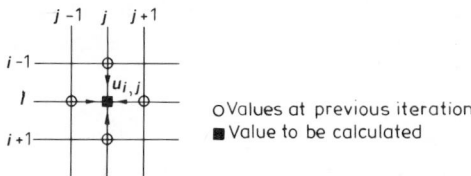

Fig. 4.2 Update for point Jacobi method

At each iteration, $u_{i,j}$ is replaced by the average value of its four adjacent neighbours with $\frac{1}{4}q_{i,j}$ subtracted. Computation is carried out in parallel as follows (see Fig. 4.2).

Allocate each grid point to a single processor in the lattice. Perform shifts in the direction indicated to bring data into the correct positions, and then update all points simultaneously. Care should be taken to keep a separate copy of grid values after each shift in order to reposition the points correctly. Provided that enough processors are available, all points on the mesh can be updated with only four termwise addition/subtraction operations.

With a little subtlety, the number of arithmetic operations can be further reduced (see Fig. 4.3):

Step 1 Add the north and west neighbours at each point (one termwise addition).

Step 2 Add the south-east neighbour to each point, in the resulting matrix, and subtract q_{ij} (one termwise addition and subtraction).

We now consider boundary conditions, of which there are basically two types: *Dirichlet* and *Neumann*. The former indicates that the values of $u(x, y)$ are specified on the boundary, and the latter that its derivatives are specified on the boundary. Both conditions can be dealt with by a suitable choice of coefficients of

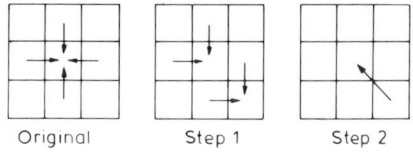

Fig. 4.3 Computation of update for point Jacobi method

the function $u^{(p)}$, which depend on the grid points. Thus, for example, instead of the constant coefficient $\frac{1}{4}$ in (4.9), we may have

$$u_{i,j}^{(p+1)} = a_{ij}u_{i-1,j}^{(p)} + b_{ij}u_{i+1,j}^{(p)} + c_{ij}u_{i,j-1}^{(p)} + d_{ij}u_{i,j+1}^{(p)} - q_{ij},$$

where the coefficients a_{ij}, b_{ij}, c_{ij}, and d_{ij} can be determined according to whether the grid points are near the boundary or not. On a parallel machine, the products can be satisfied simultaneously at each grid point by computing each product term (for example, $a_{ij}u_{i-1,j}^{(p)}$) using termwise multiplication.

Simple Dirichlet boundary conditions can be handled by an easier method. Suppose, for example, that the mesh is defined by the rectangular region

$$\{(x, y) : 1 \leqslant x \leqslant n; 1 \leqslant y \leqslant m\},$$

with boundary specified on the lines

$$\{(x, y) : x = 1, n\} \cup \{(x, y) : y = 1, m\}.$$

The interior points, forming the region $\{(x, y) : 2 \leqslant x \leqslant n - 1; 2 \leqslant y \leqslant m - 1\}$, are then updated iteratively until some convergence condition, such as $|u_{i,j}^{(p+1)} - u_{i,j}^{(p)}| < \varepsilon$ for all i and j, is satisfied. The iterative loop updates only the *interior* grid points.

(ii) *Gauss–Seidel method*

On a serial machine, convergence is about twice as slow with the Jacobi as with the Gauss–Seidel method. The Gauss–Seidel, which is a natural extension of the point Jacobi method, uses the most recent values of u on the right-hand side of (4.9) as soon as they become available. There is the added advantage of having to store only the latest value of each $u_{i,j}$. The Gauss–Seidel

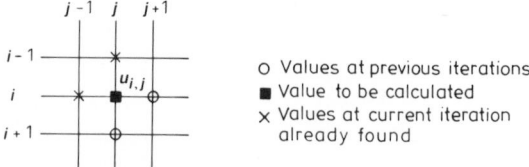

Fig. 4.4 Update for Gauss–Seidel method

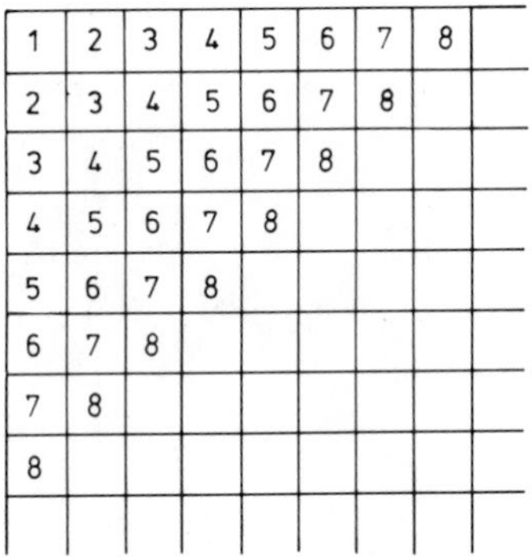

1	2	3	4	5	6	7	8	
2	3	4	5	6	7	8		
3	4	5	6	7	8			
4	5	6	7	8				
5	6	7	8					
6	7	8						
7	8							
8								

Fig. 4.5 Sequence of update in Gauss–Seidel method

iteration is defined by

$$u_{i,j}^{(p)} = \tfrac{1}{4}(u_{i,j-1}^{(p+1)} + u_{i-1,j}^{(p+1)} + u_{i,j+1}^{(p)} + u_{i+1,j}^{(p)} - q_{i,j}) \qquad (4.10)$$

(see Fig. 4.4).

We cannot immediately update all the values, as with the Jacobi method, because at each point the new value depends on two values from the previous iteration and two values from the current iteration which have already been computed. Thus the method imposes some sort of ordering, as shown in Fig. 4.5.

At the first step, point 1 is updated using the boundary condition and the initial estimate for points 2. At the second step, points 2 are updated simultaneously using the initial estimate for points 3 and the value just calculated for point 1. At the third step, points 3 are updated using current iteration values for points 2 and the initial estimate for points 4; simultaneously, we can update point 1 for the second time, because all the information is now available; and so on. Thus the first iteration sweeps across the mesh, followed in turn by the second and subsequent iterations. Rather than have mesh points at different iteration levels in the course of computation, it seems reasonable to have only two iteration levels, which are defined by the black/white† chequerboard shown in Fig. 4.6. The two levels are defined so that black points are updated first, using previous

---

† Colleagues in the U.S.A. seem to prefer the terminology red/black ordering.

Fig. 4.6 Two-level iteration for Gauss–Seidel method

white values, and white points are then updated using the latest
black values, and so on. Thus all black points are updated at one
iteration level, and all white points are updated at the next
iteration level. In Table 4.1, we make a comparison between the
point Jacobi method and the Gauss–Seidel iteration. The figures
in parenthesis indicate iteration level. After each iteration with
the Jacobi method, both the black and the white points are at the
same level.

On the other hand, after one iteration of the Gauss–Seidel
method, the black points are at level 1 (as in the Jacobi method),
but the white points have been updated and are at level 2. In the
evaluation of the update formula, the Jacobi method updates all
points but the Gauss–Seidel only half. Hence, for a square grid,

Table 4.1 Comparison of point Jacobi method and
Gauss–Seidel iteration

Iteration	1	2	3	4
Jacobi	B(1)	B(2)	B(3)	B(4)
	W(1)	W(2)	W(3)	W(4)
Gauss–Seidel	B(1)	B(3)	B(5)	B(7)
	W(2)	W(4)	W(6)	W(8)

the two methods require the same time. However, on an $n \times n$ array of processors, a $2n \times n$ problem can be solved using the Gauss–Seidel method in the same time as an $n \times n$ problem using the Jacobi method, by assigning $n \times n$ matrices for black and white points separately and updating them simultaneously.

(iii) *Successive-overrelaxation* (SOR) *method*

The successive-overrelaxation method is a variant of the Gauss–Seidel which improves the rate of convergence of the function $u(x, y)$ by adding an amount $\omega \Delta u(x, y)$, where $\Delta u(x, y)$ is the change due to the standard Gauss–Seidel iteration. The quantity ω is termed the *acceleration parameter* or the *relaxation factor*. Thus, at the $(p + 1)$th step,

$$u^{(p+1)} = u^{(p)} + \omega \Delta u, \qquad \Delta u = u_{GS}^{(p+1)} - u^{(p)}, \qquad (4.11)$$

where the subscript GS indicates Gauss–Seidel iteration. It has been shown that the process (4.11) converges if $\omega < 2$ (Varga, 1962). The use of $\omega > 1$ is termed *overrelaxation*, and $\omega < 1$ *underrelaxation* (see Fig. 4.9). We can parallelize (4.11) easily, using the black/white ordering of the Gauss–Seidel method (4.10) to produce the scheme

$$u_{i,j}^{(p+1)} = (1 - \omega)u_{i,j}^{(p)} + \tfrac{1}{4}\omega(u_{i,j-1}^{(p+1)} + u_{i-1,j}^{(p+1)} + u_{i,j+1}^{(p)} + u_{i+1,j}^{(p)} - q_{i,j}).$$

$$(4.12)$$

At the first step, $u_{i,j}^{(p+1)}$ is updated according to (4.12) for all black points, i.e. for all even $i + j$, and—at the second step—for all white points, i.e. for all odd $i + j$. Both steps consist of independent computations carried out simultaneously. The process is repeated, with a suitable choice for ω (see below), until the required degree of accuracy is achieved. It is necessary to retain only the latest values of $u_{i,j}$.

Clearly, for $\omega = 1$, the SOR method reduces to the Gauss–Seidel method.

(iv) *Line successive-overrelaxation* (LSOR) *method*

In the line successive-overrelaxation method, as the name suggests, an entire line of points is updated simultaneously. In particular, points on the column j are updated according to the

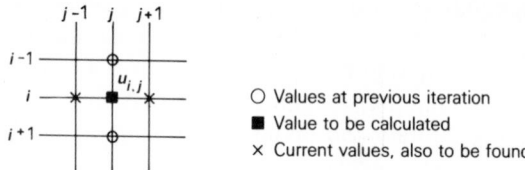

Fig. 4.7 Update for line successive-overrelaxation method

scheme

$$u_{i-1,j}^{(p+1)} - 4u_{i,j}^{(p+1)} + u_{i+1,j}^{(p+1)} = q_{i,j} - u_{i,j+1}^{(p)} - u_{i,j-1}^{(p)} \qquad (4.13)$$

(see Fig. 4.7).

The left-hand side of eqn (4.13) indicates that, for each column *j*, we need to solve a tridiagonal linear system, because each point on it is related to two neighbouring points (shown by **x** in the above diagram). We can thus treat the two-dimensional problem as a series of weakly coupled one-dimensional problems, which can be solved in parallel using cyclic odd–even reduction (Chapter 2, Section 2.6), and then introduce the coupling iteratively.

If we use the old values to update each column, we obtain the *line Jacobi* method; whereas, if we use the values just updated for the previous column, we obtain the *line Gauss–Seidel* method. As with the Gauss–Seidel method, on a parallel machine, we can adopt a two-level iteration by updating alternate columns simultaneously, as shown in Fig. 4.8. The black columns are updated using previous white column values, and then white columns are updated using the latest black values, and so on. Thus, all black columns are updated at one iteration level, and all white columns at the next iteration level.

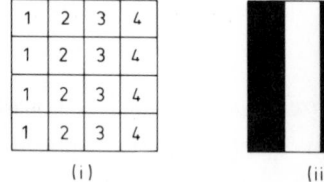

Fig. 4.8 Black/white Gauss–Seidel ordering using two iteration levels

Further, we can incorporate the *relaxation factor* ω to improve convergence:

Solve $u_{i,j}^{(p+1)}$:

$$\left.\begin{aligned} u_{i-1,j}^{(p+1)} - 4u_{i,j}^{(p+1)} + u_{i+1,j}^{(p+1)} &= q_{i,j} - u_{i,j+1}^{(p)} - u_{i,j-1}^{(p)}. \\ \textit{Overrelax:} \quad u_{i,j}^{(p+1)} &\leftarrow (1-\omega)u_{i,j}^{(p)} + \omega u_{i,j}^{(p+1)}. \end{aligned}\right\} \tag{4.14}$$

This is now referred to as the *line successive-overrelaxation* (LSOR) method. The basic steps involved in the method are as follows.

(i) *Compute* the right-hand side of the first equation in (4.14):

$$q_{i,j} - u_{i,j+1}^{(p)} - u_{i,j-1}^{(p)} = r_{i,j}^{(p)} \quad \text{(say)}$$

for each black column j.

(ii) *Solve* exactly the tridiagonal system for each black column j to give $u_{i,j}^{(p+1)}$:

$$u_{i-1,j}^{(p+1)} - 4u_{i,j}^{(p+1)} + u_{i+1,j}^{(p+1)} = r_{i,j}^{(p)}.$$

(iii) *Incorporate* overrelaxation:

$$u_{i,j}^{(p+1)} \leftarrow (1-\omega)u_{i,j}^{(p)} + \omega u_{i,j}^{(p+1)},$$

and update the black columns.

(iv) *Repeat* the above steps for the white columns.

Exact methods to determine ω do exist but may prove inordinately expensive computationally. The optimal choice is usually arrived at through experimentation. The number of iterations to convergence does not vary symmetrically with ω about the optimal value ω_{opt}, as indicated in Fig. 4.9. For the case of two-dimensional $n \times n$ discretization of the Poisson equation, with rectangular boundary,

$$\omega_{\text{opt}} = \frac{2}{1 + \sin \pi/n},$$

with a suitable choice of ω, the combination of direct and iterative methods can produce powerful schemes. However, the difficulty is that we must decide which mesh direction to treat implicitly and which explicitly. A method known as *alternating-direction implicit* has been introduced to tackle this problem. For

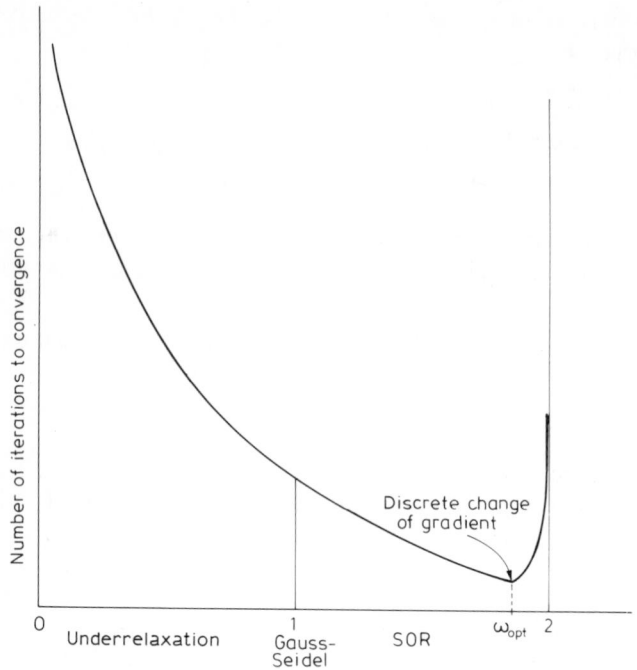

Fig. 4.9 Variation of convergence rate with overrelaxation parameter

further discussion, the reader is referred to Varga (1962) and Young (1971). Le Veque and Trefethen (1988) show that, for Poisson's equation on a rectangular region, ω_{opt} and the corresponding convergence rate may be obtained rigorously by Fourier analysis.

(v) *Block iterative schemes*

On an asynchronous MIMD system it may be preferable to use *explicit block iterative* schemes, where iterates at groups of grid points are evaluated using a direct method. This approach also reduces the number of iterations required for convergence, and is similar in nature to the LSOR method discussed earlier.

Let us consider, for example, the group of grid points given in Fig. 4.10 to evaluate the Poisson equation defined by (4.1) for $u(x, y)$, with $q_p(x, y)$ describing the sources and sinks at point p in the (x, y) plane.

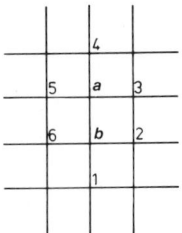

Fig. 4.10 A rectangular grid

Applying eqn (4.8) to the points a and b gives the formulae:

$$u_a = \tfrac{1}{4}(u_b + u_3 + u_4 + u_5 - q_a), \qquad u_b = \tfrac{1}{4}(u_1 + u_2 + u_a + u_6 - q_b),$$
$$(4.15)$$

which can be rearranged to give

$$u_a = \tfrac{1}{15}[4(u_3 + u_4 + u_5 - q_a) + u_1 + u_2 + u_6 - q_b], \quad (4.16)$$
$$u_b = \tfrac{1}{15}[4(u_1 + u_2 + u_6 - q_b) + u_3 + u_4 + u_5 - q_a]. \quad (4.17)$$

In the Cartesian coordinate system in terms of (i, j), we obtain

$$u_{i,j} = \tfrac{1}{15}[4(u_{i-1,j} + u_{i,j+1} + u_{i,j-1} - q_{i,j-1})$$
$$+ u_{i+1,j+1} + u_{i+2,j} + u_{i+1,j-1} - q_{i+1,j}],$$
$$u_{i+1,j} = \tfrac{1}{15}[4(u_{i+1,j+1} + u_{i+2,j} + u_{i+1,j-1} - q_{i+1,j})$$
$$+ u_{i-1,j} + u_{i,i+1} + u_{i,j-1} - q_{i,j}].$$

These equations are independent of each other, and can thus be evaluated simultaneously. Further, we can partition the mesh into 2×1 blocks, where points within each block can be updated simultaneously using *two* processors. On an array of processors, we can solve for more than one block at a time using black/white chequerboard ordering, shown in Fig. 4.6, where each square represents a 2×1 block. At the first pass, the black squares are updated simultaneously, using the white squares; at the second pass, the white squares are updated using the latest black values; and so on. In addition, we can incorporate the acceleration parameter ω, as in the case of the SOR method, and define the two-pass scheme at the $(p + 1)$th step as follows.

At the first pass:

$$u_{i,j}^{(p+1)} = u_{i,j}^{(p)} + \tfrac{1}{15}\omega(4r_{i,j}^{(p)} + r_{i+1,j}^{(p)} - 15u_{i,j}^{(p)})$$

$$u_{i+1,j}^{(p+1)} = u_{i+1,j}^{(p)} + \tfrac{1}{15}\omega(4r_{i+1,j}^{(p)} + r_{i,j}^{(p)} - 15u_{i+1,j}^{(p)}),$$

where

$$r_{i,j}^{(p)} = u_{i-1,j}^{(p)} + u_{i,j+1}^{(p)} + u_{i,j-1}^{(p)} - q_{i,j},$$

$$r_{i+1,j}^{(p)} = u_{i+1,j+1}^{(p)} + u_{i+2,j}^{(p)} + u_{i+1,j-1}^{(p)} - q_{i+1,j}.$$

At the second pass:

$$u_{i,j}^{(p+1)} = u_{i,j}^{(p)} + \tfrac{1}{15}\omega(4r_{i,j}^{(p+1)} + r_{i+1,j}^{(p+1)} - 15u_{i,j}^{(p)}),$$

$$u_{i+1,j}^{(n+1)} = u_{i+1,j} + \tfrac{1}{15}\omega(4r_{i+1,j}^{(p+1)} + r_{i,j}^{(p+1)} - 15u_{i+1,j}^{(p)}).$$

We can also derive similar formulae for larger block sizes, such as 2×2 and 3×3. Evans (1984) reports experimental results with SOR schemes using various block sizes on an asynchronous MIMD system with a moderately small number of processors (<10). As expected, by increasing the block size, the number of iterations required for convergence is decreased. On a serial machine, the number of additions and multiplications is not halved by using 2×2 block size instead of 2×1. However, provided that we can allocate four processors per 2×2 block, the number of *parallel* or *termwise* operations *is* approximately halved, and thus the choice of 2×2 block size is justified when sufficient processors are available. The use of 3×3 block size is not so effective.

Parallel algorithms for iterative methods of solving elliptic equations have been widely studied. Several authors in the 1970s (for example, Erickson 1972 and Lambiotte 1975) observed that by using the classical black/white chequerboard ordering, the Gauss–Seidel method could be carried out using two vectors of length $\tfrac{1}{2}n^2$ corresponding to the black and white points on an $n \times n$ mesh. This ordering constitutes an efficient implementation of the SOR method, but it does not extend to higher-order finite-difference or finite-element discretization, or to general elliptic equations containing mixed derivatives. It was observed, however, that black/white ordering could be extended to a 'multicolour' ordering (with the same effect as black/white ordering) for the 5-point formula given in eqn (4.10). Young (1971) was one of the first to employ 3-colour ordering, though not in the context of parallel computing. Later, in 1978,

Hackbusch used a 3-colour ordering for 9-point finite-difference formulae, and more recently Adams and Ortega (1982) generalized to p multi-colour orderings. A variety of examples are given by Adams (1982) on regular regions in two and three dimensions, where the discretization pattern is repeated systematically at each grid point. Further orderings have also been described by O'Leary (1984). In fact, there is an extensive bibliography available, and a large part of it is cited in the excellent review by Ortega and Voigt (1985).

4.5 Finite-element solution of partial differential equations

In this section, we begin with a discussion of the finite-element technique for solving partial differential equations, and then consider the applicability of parallel computation to the operations associated with it. There is much similarity between the structure of the equations resulting respectively from finite-difference and finite-element discretization. In the finite-difference method, the solution is defined at each grid point; whereas, with the finite-element method, the solution is defined everywhere using piecewise-polynomial approximation. Structural engineers have been using the finite-element approach for a fairly restricted class of problem, such as aircraft design, since the 1950s. However, it was realized only recently that it involves a classical piece of mathematics called the Ritz–Galerkin technique. More theoretical work has since been carried out, particularly using functional analysis, and the method is now gaining recognition as an important tool in a number of other areas of engineering such as stress analysis, fluid dynamics, and electromagnetism.

The finite-element method

The method is introduced by taking a two-dimensional second-order elliptic differential equation

$$
\begin{aligned}
-\Bigg[&\frac{\partial}{\partial x_1}\left(a_{11}(x)\frac{\partial}{\partial x_1}u(x)\right) + \frac{\partial}{\partial x_2}\left(a_{12}(x)\frac{\partial}{\partial x_2}u(x)\right) \\
&+ \frac{\partial}{\partial x_1}\left(a_{21}(x)\frac{\partial}{\partial x_1}u(x)\right) + \frac{\partial}{\partial x_2}\left(a_{22}(x)\frac{\partial}{\partial x_2}u(x)\right)\Bigg] = f(x)
\end{aligned}
\tag{5.1}
$$
$$
(x \in \Omega),
$$

with homogeneous Dirichlet boundary conditions given by $u(x) = 0$ $(x \in C)$, where Ω is a domain in $\mathbb{R}^2$ and is bounded by a continuous curve C.

We shall use the *Galerkin method* to find an approximate solution. First rewrite (5.1) in its *weak form* by multiplying by any function $v \in H_0^1(\Omega)$ that vanishes on the boundary C, integrate over Ω, and then transform the left-hand side of the equation by Green's theorem.

We thus require $u \in H_0^1(\Omega)$ such that

$$a(u, v) = \langle f, v \rangle \quad \text{for } v \in H_0^1(\Omega), \tag{5.2}$$

where

$$a(u, v) = \int_\Omega \sum_{i,j=1,2} a_{ij}(x) \frac{\partial}{\partial x_i} u(x) \frac{\partial}{\partial x_j} v(x) \, dx,$$

$$\langle w, v \rangle = \int_\Omega w(x) v(x) \, dx.$$

Here $H_0^1(\Omega)$ is the subspace of the *Sobolev space* $H^1(\Omega)$ defined by

$$H_0^1(\Omega) = \{ v \in H^1(\Omega) : v = 0 \text{ on } C \},$$

with the same norm as for $H^1(\Omega)$. The *Sobolev space* $H^s(\Omega)$ of order s is the space of functions w defined on $\Omega \subseteq \mathbb{R}^n$, such that w and all its (mixed) partial derivatives up to order s exist throughout Ω and each have finite $L_2(\Omega)$ norm: for example, if $s = 1$, this implies that the Sobolev norm

$$\|w\|_{H^1(\Omega)} = \left\{ \int_\Omega \left[w^2(x) + \sum_{i=1}^n \left(\frac{\partial w}{\partial x_i} \right)^2 (x) \right] dx \right\}^{\frac{1}{2}}$$

is finite.

Returning to eqn (5.1), if $a_{ij} = \delta_{ij}$ (where $\delta_{..}$ is the Kronecker delta function) we obtain the Poisson equation

$$-\left(\frac{\partial^2 u}{\partial x^2} + \frac{\partial^2 u}{\partial y^2} \right)(x, y) = f(x, y) \quad \text{for } (x, y) \in \Omega,$$

$$u(x, y) = 0 \quad \text{for } (x, y) \in C. \tag{5.3}$$

From the associated weak form, we seek a function $u \in H_0^1(\Omega)$

such that

$$\int_\Omega \nabla u(x, y) \cdot \nabla v(x, y) \, \mathrm{d}x = \int_\Omega f(x, y)v(x, y) \, \mathrm{d}x \qquad (5.4)$$

for all $v \in H_0^1(\Omega)$, where $\nabla u = \boldsymbol{i} \, \partial u/\partial x + \boldsymbol{j} \, \partial u/\partial y$ and $\nabla v = \boldsymbol{i} \, \partial v/\partial x + \boldsymbol{j} \, \partial v/\partial y$. We now apply the Galerkin method to the weak form (5.4) to find an approximate solution for u. The domain Ω is divided into a set of elements using *simplicial decomposition* (for example, such that the intersection of any two triangles of the decomposition is an edge of each of them). Typically, we choose either 3-node triangular or 4-node or 8-node quadrilateral elements in the two-dimensional case, and 8-node or 20-node brick elements in the three-dimensional case. A finite-dimensional subspace of functions $v^h \in H_0^1(\Omega)$ is chosen which, in general, consists of piecewise polynomials defined over each element Ω^h. Then we look for an approximate solution u^h such that

$$\int_{\Omega^h} \nabla u^h(\boldsymbol{x}) \cdot \nabla v^h(\boldsymbol{x}) \, \mathrm{d}x = \int_{\Omega^h} f(\boldsymbol{x})v^h(\boldsymbol{x}) \, \mathrm{d}x. \qquad (5.5)$$

The functions u^h and v^h are normally chosen from the same subspace, but this is not essential. If we choose u^h and v^h of the form

$$u^h(x, y) = \sum_{j=1}^n \alpha_j \phi_j(x, y), \qquad v^h(x, y) = \sum_{i=1}^n \beta_i \phi_i(x, y),$$

then (5.5) becomes

$$\sum_{i,j} \beta_i \alpha_j \int_{\Omega^h} \nabla \phi_j(x) \cdot \nabla \phi_i(x) \, \mathrm{d}x = \sum_i \beta_i \int_{\Omega^h} f(x)\phi_i(x) \, \mathrm{d}x. \qquad (5.6)$$

If we define

$$k_{ij} = \int_{\Omega^h} \nabla \phi_j(x) \cdot \nabla \phi_i(x) \, \mathrm{d}x, \qquad f_i = \int_{\Omega^h} f(x)\phi_i(x) \, \mathrm{d}x,$$

then (5.6) becomes

$$\sum_{i,j} \beta_i(\alpha_j k_{ij} - f_i) = 0,$$

which must be true for all β_i, and we thus obtain the following symmetric and positive definite system of linear equations for α:

$$\sum_j k_{ij}\alpha_j = f_i. \tag{5.7}$$

The matrix $[k_{ij}]$ will be sparse because of the choice of the *basis* or *shape* functions ϕ_i. In the case of triangular elements, the values of k_{ij} will be zero if node i and node j are not on a common finite element. If the partial differential equations are linear, we obtain a linear set of equations which are solved approximately using one of a number of techniques, the choice being governed largely by the pattern of zero elements in the matrix $[k_{ij}]$. The finite-element method is discussed in its broad outlines by several authors, for example, Strang and Fix (1973), Zienkiewicz (1977), Mitchell and Wait (1977), and Livesley (1983).

Machine architecture, problem size and complexity, and algorithm design

As seen in Chapters 1 and 2, the various existing parallel computers differ with respect to the number of processing elements, the type of routeing network connecting them, the size of the core memory, whether each processor has its own local memory and/or shared global memory, and whether arithmetic is done in hardware at bit level or word level. Further, there are fundamental differences between the vector processor, the SIMD system (single instruction stream, multiple data stream), and the MIMD system (multiple instruction stream, multiple data stream), and some recent designs are hybrids incorporating elements of some or all of the above systems. In principle, different phases of the finite-element process may require different parallel machine architecture for optimal efficiency; moreover, problem size and complexity, architecture, and algorithm design are interrelated, and there is a wide range of problem sizes occurring in practice. In the following discussion, we shall assume that sufficient processors are available which have SIMD characteristics with the routeing network forming a lattice (Chapter 1, Fig. 7.9). This is, however, unlikely, and the programmer may well be obliged to resort to partitioning the problem into smaller units using the slicing or crinkling schemes discussed in Chapter 2. The emphasis here will be on the exploitation of parallelism in each of the three phases: (a) computation of the local system of equations, (b) assembly into a global matrix, and (c) solution of the resulting linear system.

(a) *Computation of the local system of equations*

The operations involved in calculating the local system of equations for the different elements are independent, and can therefore be done simultaneously.

In a serial computation, each element is taken in turn, and the element calculation operations (for example, a numerical integration at each Gauss point) are carried out within the loop. On a lattice of processors, the loop goes round p times, where p is the number of Gauss points for each element; in the case of the two-dimensional Poisson equation with 8-node quadrilateral elements, p is 9. Within the loop, a particular Gauss point is considered for all the elements simultaneously. Next time round the loop, another Gauss point is considered, again for all the elements simultaneously, and so on. Figure 5.1 contrasts the serial and parallel schemes.

One advantage of the lattice is that the processors are connected to their nearest four neighbours, which provides a natural means of communication; whereas, with a serial algorithm, it is necessary to keep a separate record of the neighbouring nodes and to index these using a table look-up.

The following operations can be carried out in parallel, simultaneously for all the elements:

(i) determination of the basis functions, and their local derivatives,

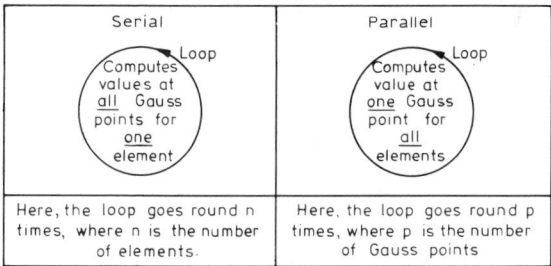

Serial	Parallel
Loop Computes values at <u>all</u> Gauss points for <u>one</u> element	**Loop** Computes value at <u>one</u> Gauss point for <u>all</u> elements
Here, the loop goes round n times, where n is the number of elements.	Here, the loop goes round p times, where p is the number of Gauss points

Fig. 5.1 Exploitation of parallelism for computation of the local system of equations

   (ii)   computation of the Jacobian, and its inverse,

   (iii)   computation of global coordinates in terms of local derivatives,

   (iv)   incorporation of source strengths,

   (v)   computation of the local system of equations.

Thus phase (a) is highly parallel, particularly for large 3-D problems.

(b)   *Assembly into a global matrix*

A number of authors have studied the translation of serial algorithms and found that, although a wide variety of element configurations can be accommodated, a considerable organizational cost is incurred, and only minor utilization of parallelism is feasible. Here we describe a technique that exploits the fixed geometry of the lattice, with nearest-neighbour connectivity of processors; although the method restricts the element configuration, a considerable speed-up can be achieved—indeed, relative to the other phases, the time for global assembly is found to be negligible.

We assume that the mesh is divided into rectangular elements so as to fit on to the lattice, with a maximum of $(n - 1) \times (n - 1)$ elements so that each element and/or node maps onto one processor. For a larger mesh size, the standard technique of partitioning the region into sets of $(n - 1) \times (n - 1)$ blocks would be employed. This mapping makes it easy to utilize the connectivity of the processing elements. For example, for a 4-point Gaussian quadrature, the values of the Gauss points for all the elements are carefully stored so that, in the summation process, the relevant values are placed in adjacent processors (as far as possible) and hence are directly accessible. For more Gauss points, or for a higher number of nodes per element, the same technique may be employed. However, nonutilization of some of the processing elements may become necessary, leading to a reduction in the size of problem that can be solved completely in parallel: for example, for 8-node elements, we can arrange the computation so that approximately a $\frac{1}{2}n \times \frac{1}{2}n$ mesh can be processed simultaneously on an $n \times n$ lattice of processors.

To illustrate how parallelism can be exploited, we discuss assembly for Laplace's equation in the simplified case of a region

Fig. 5.2 (a) Plane rectangular region, (b) local coordinate with standard quadrilateral element

with 9 rectangular elements (Fig. 5.2) and a 4×4 lattice (for higher-order elements and large 3-D problems, the parallelism can generally be identified in a similar way). The entries of any 4×4 matrix are computed and stored with the corresponding processing elements.

Suppose we have computed the local system (4×4) for each element, and the entries comprising the coefficient matrices are stored as follows.

$$A11 = \begin{bmatrix} (11)^1 & (11)^2 & (11)^3 & 0 \\ (11)^4 & (11)^5 & (11)^6 & 0 \\ (11)^7 & (11)^8 & (11)^9 & 0 \\ 0 & 0 & 0 & 0 \end{bmatrix},$$

$$A22 = \begin{bmatrix} (22)^1 & (22)^2 & (22)^3 & 0 \\ (22)^4 & (22)^5 & (22)^6 & 0 \\ (22)^7 & (22)^8 & (22)^9 & 0 \\ 0 & 0 & 0 & 0 \end{bmatrix},$$

$$A33 = \begin{bmatrix} (33)^1 & (33)^2 & (33)^3 & 0 \\ (33)^4 & (33)^5 & (33)^6 & 0 \\ (33)^7 & (33)^8 & (33)^9 & 0 \\ 0 & 0 & 0 & 0 \end{bmatrix},$$

$$A44 = \begin{bmatrix} (44)^1 & (44)^2 & (44)^3 & 0 \\ (44)^4 & (44)^5 & (44)^6 & 0 \\ (44)^7 & (44)^8 & (44)^9 & 0 \\ 0 & 0 & 0 & 0 \end{bmatrix},$$

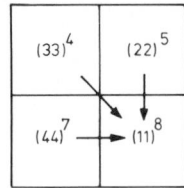

Fig. 5.3 Shifts required for alignment of entries

and similarly for all $i,j = 1,2,3,4$. Here $(ij)^p$ denotes the (i, j) entry in the coefficient matrix for the element p. If we now consider an entry, for example, in position $(10, 10)$ in the global matrix, we see from Fig. 5.2 that it is formed as the sum of the numbers

$$(11)^8, \quad (22)^5, \quad (33)^4, \quad (44)^7.$$

By superimposing the four matrices, we can see their relative positions in $A11$ (Fig. 5.3). The arrows indicate the shifts required to the matrices $A22$, $A33$, and $A44$, so that the other three entries can be aligned with $(11)^8$.

The calculation of other diagonal entries on the global matrix involves shifts in exactly the same direction, and thus all shifts can be performed in parallel. The careful alignment of elements in their proper locations allows the computation to be carried out simultaneously using termwise operations as listed in Section 2.2 of Chapter 2. The rest of the bands in the global matrix can be computed in a similar way.

On a coarse-grain MIMD machine the assembly may be carried out as follows. Suppose we have a three-processor system, and 8 three-noded triangular elements as shown in Fig. 5.4.

With d unknowns at each node, the resulting stiffness matrix will be of dimensions $9d \times 9d$. Each group of three nodes is allocated a single processor; for example, nodes 4, 5, and 7 are allocated processor P. During the computation, each processor will calculate exactly $3d$ unknowns. Thus a $3d \times 3d$ linear system $Kx = b$ must be solved. But first the coefficients in the matrix K need to be computed; in the case of processor P, this is done as follows.

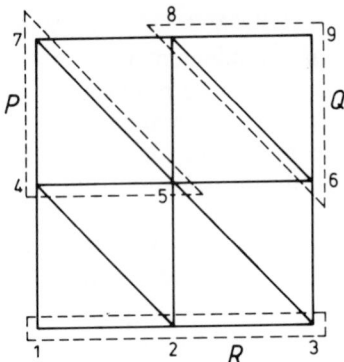

Fig. 5.4 Finite-element discretization with 8 three-noded triangular elements

The coefficients required by processor P for the solution of unknowns at nodes 4, 5, and 7 must be calculated either by P or by processors R or Q and communicated to P. If we label the nodes 4, 5, and 7 as *interior nodes* to processor P, and other nodes which share common finite elements as *exterior nodes* (residing in other processors), then the resulting data communication is shown in Table 5.1, where x represents a $d \times d$ data matrix.

The amount of storage per processor depends on the number of nodes assigned to P, the number of equations at each node, and the number of nodes that share common finite elements with P's nodes. Further, for a direct method of solution such as the Choleski factorization, extra storage in processor P must be reserved for the coefficients that will 'fill in' the band. For a detailed description see Adams (1982).

Table 5.1 Data communication required to compute coefficients of K in processor P

	Interior nodes			Exterior nodes				
	4	5	7	1	2	3	6	8
4	x	x	x	x	x			
5	x	x	x		x	x	x	x
7	x	x	x					x

(c) *Solution process: conjugate-gradient method*

In the conventional implementation of the finite-element method, the solution procedure requires the global matrix either to be assembled from the local systems beforehand, in the case of the elimination–decomposition algorithm, or to be assembled gradually as the solution proceeds, in the frontal-solution algorithm. Both these techniques suffer from 'fill-in', which can be serious for a banded matrix which is often initially very sparse. Iterative techniques do not involve 'fill-in', and the conjugate-gradient (CG) method can be adapted successfully to parallel computation. In fact, with this method, there is no need to assemble the global matrix if it is carefully coded, and a significant amount of parallelism can be introduced in the matrix by vector multiplication, which forms the core of the computation. To reduce the number of steps, some kind of preconditioning is necessary: for example, the transformation of $Ax = b$ into $A'x' = b'$ with a much smaller condition number in order to speed up the convergence (see below).

Implementation of the CG *method*

The CG method without preconditioning is as follows. To solve $Ax = b$, where A is a positive definite symmetric $n \times n$ matrix:

Set an initial approximation vector x_0,
calculate the initial residual $r_0 = b - Ax_0$,
set the initial search direction $p_0 = r_0$;
then, for $i = 0, 1, \ldots,$

(a) calculate the coefficient $\alpha_i = p_i^T r_i / p_i^T A p_i$,
(b) set the new estimate $x_{i+1} = x_i + \alpha_i p_i$,
(c) evaluate the new residual $r_{i+1} = r_i - \alpha_i A p_i$,
(d) calculate the coefficient $\beta_i = -r_{i+1} A p_i / p_i^T A p_i$,
(e) determine the new direction $p_{i+1} = r_{i+1} + \beta_i p_i$,
        continue until either r_i or p_i is zero.

The values α_i ensure that the A^{-1}-norm $\|r_{i+1}\|_{A^{-1}}$ of the residual r_{i+1} is a minimum along the line $\{x_i + \alpha p_i : \alpha \in \mathbb{R}\}$ at $\alpha = \alpha_i$, where $\|x\|_{A^{-1}} = (x^T A^{-1} x)^{\frac{1}{2}}$. Also

$$r_i^T p_j = 0 \quad (i > j).$$

The values β_i ensure the conjugacy condition

$$p_i^T A p_j = 0 \quad (i \neq j),$$

and it can easily be shown that $r_i^T r_j = 0$ $(i \neq j)$.

The following observations identify some parallel features of the CG method.

Line (a) uses the special-purpose parallel matrix-by-vector multiplier: see Parkinson (1981) for multiplication of unstructured sparse matrices, and Barlow *et al.* (1984) for multiplication of structured sparse matrices. The inner products in the numerator and denominator are calculated in parallel.

Line (b) updates all the elements of x simultaneously using termwise operations. The scalar α_i is mapped onto a vector with each entry equal to α_i, and then multiplied with p_i using termwise multiplication. The resulting vector is then added to the vector x_i in parallel using termwise addition.

The rest of the steps, lines (c)–(e), are evaluated similarly.

Preconditioning the CG *method*

Various techniques for preconditioning the CG method have been under investigation, for vector and parallel processors as well as for serial machines (Jennings 1977, 1982). These include Jacobi, incomplete Choleski, polynomial approximation, block Jacobi, and block cyclic reduction. Their effectiveness varies according to the characteristics of the stiffness matrix. Block methods appear to be more attractive for both vector and parallel processors.

Kowalik and Kumar (1982) examine the use of block Jacobi preconditioning on the Denelcor HEP MIMD computer and demonstrate significant improvement. The polynomial-approximation method also has a high degree of intrinsic parallelism, of a nature suitable for vector processing. More details on preconditioning the CG method for vector and parallel computers are given by Jordan (1984) and Lai and Liddell (1988), and a review of preconditional CG methods for solving systems of algebraic equations is contained in Fletcher (1986) and Jacobs (1980). For further information on the parallel implementation of the CG method, the reader might consult Dixon and Duxbury (1985) and Duff (1987).

5 The symmetric eigenvalue problem: Jacobi method

5.1 Introduction

In the ensuing chapter we shall examine the Jacobi method. We seek scalars λ, and corresponding vectors x, such that

$$Ax = \lambda x, \qquad x \neq 0, \tag{1.1}$$

where A is an $n \times n$ symmetric matrix. Each such λ is an *eigenvalue* of A, with x an *eigenvector* of A corresponding to λ. Clearly (1.1) holds if and only if $A - \lambda I$ is singular. Thus the eigenvalues are precisely the roots of the *characteristic equation*

$$\det(A - \lambda I) = 0, \tag{1.2}$$

denoted by $\lambda_1, \ldots, \lambda_n$, with the multiplicity m_λ of any eigenvalue λ equal to that of the factor $\lambda - x$ of the x-polynomial $\det(A - xI)$. Since A is symmetric, all the eigenvalues are real. For each eigenvalue λ, the corresponding eigenvectors constitute a vector subspace of $\mathbb{R}^n$ of dimension m_λ. The eigenvectors of A span $\mathbb{R}^n$, and from them is chosen a linearly independent set $\{x, \ldots, x_n\}$ of representatives such that x_i corresponds to λ_i, for $i = 1, \ldots, n$.

Certain features of the Jacobi method make it particularly well suited to a multiprocessor system. This is one reason for choosing to focus on it. Another is to illustrate the fact that parallel architectures may require us to reconsider outdated serial methods. We first give a brief resumé of the contents of this chapter.

The Jacobi method was first proposed in 1846, but, although it is a well-known technique for computing eigensolutions, it was not until the development of automatic machines that it could be fully appreciated. In this century, it has been studied by many research workers, including Henrici, Forsythe, and Wilkinson, and the general consensus has been that its virtues are simplicity and reliability. In more recent years, however, it has been somewhat eclipsed by other methods, such as the QR algorithm (1961), which proved more economical and offered the same reliability on serial machines.

Due to the emergence of parallel computers, the Jacobi method has once again come to the forefront of our attention. We have concentrated on examining the variants of the cyclic Jacobi scheme, performing $\frac{1}{2}n$ rotations simultaneously (where n is an even number representing the order of the matrix). The results are promising, and encourage us in our intention to stimulate interest in parallel cyclic Jacobi schemes by exploring developments both theoretical and practical.

The initial sections (5.1 to 5.3) sketch in the background to the Jacobi method. Section 5.4 shows how several orthogonal transformations can be carried out simultaneously, and introduces the concept of the compound rotational matrix, which consists of a set of independent rotations. These ideas are then adapted in the formulation of the parallel method (Section 5.6), together with a discussion of convergence properties (Section 5.9). In Section 5.7 we provide a general framework for describing mobile schemes which minimize data movement. A variant of the cyclic Jacobi method is proposed in Section 5.8. The later sections discuss in detail the application of approximate formulae to rotational angles and the application of the perturbation method. The chapter concludes with the exposition of a simple extension of the parallel Jacobi method for the solution of the generalized eigenvalue problem.

5.2 The classical method

A real $n \times n$ symmetric matrix A is reduced to diagonal form by application of plane rotations. The element $a_{pq}^{(k)}$ is annihilated by applying a plane rotation in the (p, q)-plane at an angle $\alpha_{pq}^{(k)}$, according to the scheme (Wilkinson 1965):

$$A^{(1)} = A, \qquad A^{(k+1)} = J^{(k)}A^{(k)}J^{(k)\mathsf{T}} \quad (k = 1, 2,...),$$

where $J^{(k)}$ is a block-diagonal matrix such that

$$J_{(i,j)}^{(k)} = \delta_{ij} \quad (i,j \neq p,q), \qquad \delta_{ij} = \begin{cases} 1 & \text{if } i = j, \\ 0 & \text{if } i \neq j, \end{cases}$$

$$J_{(p,q)}^{(k)} = s = -J_{(q,p)}^{(k)}, \qquad J_{(p,p)}^{(k)} = c = J_{(q,q)}^{(k)},$$

with $s = \sin \alpha_{pq}^{(k)}$ and $c = \cos \alpha_{pq}^{(k)}$. In the classical scheme, the element $a_{pq}^{(k)}$ to be annihilated is chosen so that it corresponds to the numerically largest off-diagonal element of A:

$$|a_{pq}^{(k)}| = \max_{i \neq j} |a_{ij}^{(k)}|.$$

$A^{(k+1)}$ differs from $A^{(k)}$ only in rows and columns p and q, and the modified values are as follows.

$$\left. \begin{aligned} a_{ip}^{(k+1)} &= a_{ip}^{(k)}c + a_{iq}^{(k)}s = a_{pi}^{(k+1)} \\ a_{iq}^{(k+1)} &= -a_{ip}^{(k)}s + a_{iq}^{(k)}c = a_{qi}^{(k+1)} \end{aligned} \right\} \quad (i \neq p, q), \qquad (2.1)$$

$$\left. \begin{aligned} a_{pp}^{(k+1)} &= a_{pp}^{(k)}c^2 + 2a_{pq}^{(k)}cs + a_{qq}^{(k)}s^2 \\ a_{qq}^{(k+1)} &= a_{pp}^{(k)}s^2 - 2a_{pq}^{(k)}cs + a_{qq}^{(k)}c^2 \end{aligned} \right\}, \qquad (2.2)$$

$$\left. \begin{aligned} a_{pq}^{(k+1)} &= (a_{qq}^{(k)} - a_{pp}^{(k)})cs + a_{pq}^{(k)}(c^2 - s^2) \\ a_{qp}^{(k+1)} &= a_{pq}^{(k+1)} \end{aligned} \right\}. \qquad (2.3)$$

Since we require $a_{pq}^{(k+1)}$ to be zero, from (2.3) we have

$$\tan 2\alpha_{pq}^{(k)} = 2a_{pq}^{(k)}/(a_{pp}^{(k)} - a_{qq}^{(k)}),$$

i.e.

$$\alpha_{pq}^{(k)} = \tfrac{1}{2} \tan^{-1}[2a_{pq}^{(k)}/(a_{pp}^{(k)} - a_{qq}^{(k)})], \qquad (2.4)$$

where $\alpha_{pq}^{(k)}$ is chosen to lie in the range

$$-\tfrac{1}{4}\pi < \alpha_{pq}^{(k)} \leq \tfrac{1}{4}\pi.$$

If $a_{pp}^{(k)} = a_{qq}^{(k)}$, then $\alpha_{pq}^{(k)}$ is set to $\pm\tfrac{1}{4}\pi$, according to the sign of $a_{pq}^{(k)} \neq 0$. The sequence of matrices $A^{(k)}$ is produced such that (Jacobi 1846)

$\lim_{k \to \infty} A^{(k)} = D$, a diagonal matrix representing the eigenvalues of A,

$\lim_{k \to \infty} J^{(1)\mathsf{T}} J^{(2)\mathsf{T}} \cdots J^{(k)\mathsf{T}} = Q$, an orthogonal matrix representing the eigenvectors of A.

On a serial machine, the search for the largest element is time-consuming. It is therefore customary to choose the pivots in some simple predefined order. This leads to a variant of the classical method known as the cyclic Jacobi scheme.

5.3 Cyclic schemes

In a cyclic Jacobi scheme, each sweep consists of $N = \frac{1}{2}n(n-1)$ consecutive elements of the array $((p, q) : 1 \leqslant p < q \leqslant n)$, where each pair (p, q) occurs exactly once. Goldstine defined two such schemes which have since been commonly in use (Gregory 1953). These are known as the row-cyclic and column-cyclic methods (sometimes referred to as special cyclic orderings).

The row-cyclic method is defined as:

$$(p_0, q_0) = (1, 2),$$

$$(p_{k+1}, q_{k+1}) = \begin{cases} (p_k, q_k + 1) & \text{if } p_k < n - 1 \text{ and } q_k < n, \\ (p_k + 1, p_k + 2) & \text{if } p_k < n - 1 \text{ and } q_k = n, \\ (1, 2) & \text{if } p_k = n - 1 \text{ and } q_k = n. \end{cases}$$

The column-cyclic method is defined as:

$$(p_0, q_0) = (1, 2),$$

$$(p_{k+1}, q_{k+1}) = \begin{cases} (p_k + 1, q_k) & \text{if } p_k < q_k - 1 \text{ and } q_k \leqslant n, \\ (1, q_k + 1) & \text{if } p_k = q_k - 1 \text{ and } q_k < n, \\ (1, 2) & \text{if } p_k = n - 1 \text{ and } q_k = n. \end{cases}$$

After completing a sweep the cycle is repeated, and typically, five or six sweeps have to be applied before A converges to the diagonal form to the required degree of accuracy.

Although the above methods were used extensively, the question of convergence was not settled until 1958 when Henrici proved that the ultimate rate of convergence is quadratic. The convergence properties of cyclic schemes are discussed in detail in Section 5.9.

A slight modification to the cyclic scheme, known as the threshold method, skips a rotation at the kth step if the pivot element $|a_{pq}^{(k)}|$ is small compared to the general level of the off-diagonal elements of A. The reason for this is that little progress is made in performing such a rotation. The tolerance level δ_{tol} may be defined as $\delta_{tol} = 10^{-s}d$, where

$$d = \sum_{\substack{i,j=1 \\ i \neq j}}^{n} a_{ij}^{(k)2},$$

so that a rotation is skipped if $|a_{pq}| < 10^{-s}d$. Typical values of s

might be 2, 4, and 6 for the first three sweeps, and infinity for the rest. This modification results in a modest improvement in speed. Further, as Wilkinson (1965) shows, there can only be a finite number of iterations corresponding to any given threshold value, so the process is guaranteed to terminate.

As each Jacobi rotation costs two square roots and $4n$ multiplications, the cost of a single sweep would be $2n^3 + O(n^2)$ multiplications and $O(n^2)$ square roots. More detail is given by Rutishauser (1971).

5.4 Simultaneous application of rotations

For the serial scheme, as seen in Section 5.2, each plane rotation affects row p, row q, column p, and column q of A, annihilating the (p, q)th element of A (and its symmetric counterpart). This feature of the Jacobi algorithm can be exploited in order to annihilate more than one element at a time.

We shall assume without loss of generality that A is of even order and take a matrix of order 6 to illustrate the idea.

Suppose we choose the pivots $(1, 2)$ $(3, 4)$, and $(5, 6)$ of A. In calculating the angles $\alpha_{pq}^{(k)}$, the following triples of elements are needed: (a_{11}, a_{22}, a_{12}) for $\alpha_{12}^{(k)}$, (a_{33}, a_{44}, a_{34}) for $\alpha_{34}^{(k)}$, and (a_{55}, a_{66}, a_{56}) for $\alpha_{56}^{(k)}$, according to eqn (2.4). Thus we calculate (c_i, s_i) $(i = 1, 2, 3)$, and construct the plane rotation matrix J as follows:

$$\begin{bmatrix} c_1 & s_1 & 0 & 0 & 0 & 0 \\ -s_1 & c_1 & 0 & 0 & 0 & 0 \\ 0 & 0 & c_2 & s_2 & 0 & 0 \\ 0 & 0 & -s_2 & c_2 & 0 & 0 \\ 0 & 0 & 0 & 0 & c_3 & s_3 \\ 0 & 0 & 0 & 0 & -s_3 & c_3 \end{bmatrix}$$

Fig. 4.1 A typical compound rotation matrix

We call this matrix J a *compound rotation matrix* as opposed to the single plane rotation used in the serial case. It is an orthogonal matrix, so we may construct $A^{(2)}$ by applying the orthogonal similarity transformation $A^{(2)} = JA^{(1)}J^\mathsf{T}$ with $A^{(1)} = A$

in order to annihilate the elements a_{12}, a_{34}, and a_{56} simultaneously (Fig. 4.1). The resultant matrix $A^{(2)}$ would be identical to the matrix obtained in a serial scheme if the elements a_{12}, a_{34}, and a_{56} were annihilated successively in *any order*.

From Fig. 4.1, we see that the rotations chosen are disjoint: two rotations R_{pq} and R_{rs} are said to be *disjoint* if the indices p, q, r, and s are all distinct. The row-cyclic and column-cyclic orderings given for serial computation in Section 4.3 are unsuitable because, in any row or column, the indices are not distinct.

In other words, for $m \geq 2$, a set of rotations $R_i = R_{p_i, q_i}$ ($i = 1, \ldots, m$) in which $\{p_i, q_i\} \cap \{p_j, q_j\} = \varnothing$ ($i \neq j$) is called *disjoint*; the corresponding matrices commute, and application of the *compound rotation matrix* $J = \prod_{i=1}^{m} R_i$ annihilates m elements of A by means of m rotations which can all be performed simultaneously. We write

$$\text{Piv } J = \{(p_i, q_i) : i = 1, \ldots, m\}.$$

Since each rotation affects two rows (and columns) of A, the maximum number of rotations that can be performed simultaneously is clearly $\lfloor \tfrac{1}{2}n \rfloor$, and a sweep consists of $2\lceil \tfrac{1}{2}n \rceil - 1$ steps compared with $\tfrac{1}{2}n(n-1)$ in the serial case.

The Jacobi method in which compound rotations are performed at each stage is termed the *parallel Jacobi method*.

Now define

$$e(n) = \begin{cases} n & \text{if } n \text{ is even,} \\ n-1 & \text{if } n \text{ is odd,} \end{cases} \qquad o(n) = \begin{cases} n-1 & \text{if } n \text{ is even,} \\ n & \text{if } n \text{ is odd,} \end{cases}$$

$$Z = \{(p, q) : 1 \leq p < q \leq n\}.$$

Let $Z_k \subseteq Z$ ($k = 1, \ldots, o(n)$) be such that

$$\left. \begin{array}{l} \{Z_1, \ldots, Z_{o(n)}\} \text{ partitions } Z, \\ |Z_k| = \tfrac{1}{2}e(n) \quad (k = 1, \ldots, o(n)). \end{array} \right\} \tag{4.1}$$

Then we can annihilate simultaneously the $\tfrac{1}{2}e(n)$ elements of $A^{(k)}$ indexed by the elements of Z_k, using $\tfrac{1}{2}e(n)$ disjoint rotations. Thus, the parallel Jacobi method proceeds as follows. A single sweep of the method is defined by

$$A^{(1)} = A, \qquad A^{(k+1)} = J^{(k)}A^{(k)}J^{(k)\mathsf{T}} \quad (k = 1, \ldots, o(n)),$$

where, for all $(p, q) \in Z_k$, we have

$$J^{(k)}(p, q) = -J^{(k)}(q, p) = s$$
$$J^{(k)}(p, p) = J^{(k)}(q, q) = c$$
$$J^{(k)}(i, j) = \delta_{ij} \quad \text{if} \quad \{i, j\} \not\subseteq \{p, q\}.$$

It is necessary to divide a sweep into $o(n)$ blocks of $\frac{1}{2}e(n)$ so that each block consists of $\frac{1}{2}e(n)$ disjoint rotations and no rotation is repeated within a sweep. After a sweep has been completed, the cycle is repeated with the initial A replaced by the latest $A^{(k+1)}$ computed in the previous sweep. This process is repeated until A converges to the diagonal form to the required degree of accuracy. The exact form of the method is determined by the choice of $(Z_1, \ldots, Z_{o(n)})$ satisfying (4.1). It will be seen in the next section that the choice of such an ordering has an interesting history.

5.5 Kirkman's schoolgirls' problem

The literature contains a number of different methods for determining a sequence of $\lfloor \frac{1}{2}n \rfloor$ pairs specifying the pivot elements a_{pq} that are to be simultaneously annihilated at successive stages. Indeed, the specification of such sequences may be said to date back to 'Kirkman's schoolgirls' problem', first posed in *The Lady's and Gentleman's Diary* in 1850. W. W. Ball presented the problem as follows: 'A school mistress was in the habit of taking her girls for a daily walk. The girls were fifteen in number and were arranged in five rows of three each, so that each girl might have two companions. The problem is to dispose them so that, for seven consecutive days, no girl will walk with any of her school fellows in any triplet more than once.' With $n = 15$, the problem has more than 1.5×10^{10} solutions.

A more general problem, which was the subject of substantial research during the nineteenth century, may be expressed as follows (Brodlie 1973): 'For $l \equiv 1 \pmod{m - 1}$, find an arrangement of lm girls walking in rows of m abreast so that, over $(lm - 1)/(m - 1)$ consecutive days, no girl walks with any other more than once'. It will be seen that our problem is a special case of this, when $l = \frac{1}{2}n$ (with n even) and $m = 2$. We seek a partitioning of the $\frac{1}{2}n(n - 1)$ pairs of integers (i, j) $(1 \leq i < j \leq n)$

into $n - 1$ groupings of $\frac{1}{2}n$ pairs such that each integer is represented once (and only once) in each grouping; for example, with $n = 4$, we have the three groupings

$$\{(1, 2), (3, 4)\}, \qquad \{(1, 3), (2, 4)\}, \qquad \{(1, 4), (2, 3)\}.$$

For an interesting discussion, see Ball and Coxeter (1939).

In Sections 5.6 and 5.7, we examine several such orderings and their implementation. Orderings of this type are also discussed in Chapter 3 in the context of complete parallel sorting schemes. In the meantime, there are still a number of devoted enthusiasts working on possible solutions to the schoolgirl problem in connection with parallel computing.

5.6 Design of parallel algorithms

A number of decisions which must be taken in the development of parallel algorithms are irrelevant, or at least relatively insignificant, in the development of serial algorithms: for instance, interprocessor communication, synchronization, and other 'housekeeping' activities. These design considerations have a strong bearing on the overall speed of the algorithm, which, if it is poorly designed, will not take full advantage of parallelism.

Several authors have constructed annihilation orderings. In the first instance, we will use one of Sameh's schemes (1971). Define $m = \lfloor \frac{1}{2}(n + 1) \rfloor$, where n is the order of the matrix and $\lfloor x \rfloor$ is the greatest integer less than or equal to x. A sweep therefore consists of $2m - 1$ compound rotations.

The subsets Z_k ($k = 1, \ldots, 2m - 1$) in (4.1), giving the coordinates of the disjoint rotations (p, q), may be chosen as follows.

(a) For $k = 1, \ldots, m = 1$ and for $q = m - k + 1, \ldots, n - k$:

$$p = \begin{cases} 2m - 2k + 1 - q & \text{if } m - k + 1 \leq q \leq 2m - 2k, \\ 4m - 2k - q & \text{if } 2m - 2k < q < 2m - k, \\ n & \text{if } 2m - k \leq q \leq n - k. \end{cases}$$

(b) For $k = m, \ldots, 2m - 1$ and for $q = 4m - n, \ldots, 3m - k - 1$:

$$p = \begin{cases} n & \text{if } 4m - n \leq q \leq 2m - k, \\ 4m - 2k - q & \text{if } 2m - k < q < 4m - 2k, \\ 6m - 2k - 1 - q & \text{if } 4m - 2k \leq q \leq 3m - k - 1. \end{cases}$$

Once each of the subsets Z_k is found, the transformation

$$A^{(k+1)} = J^{(k)} A^{(k)} J^{(k)\mathsf{T}}$$

is carried out. It seems ironic that the process of transforming the serial method into a parallel one implicitly involves a serial description.

In the organizational details which follow, we consider three possible improvements: (i) to find a parallel method for generating the position of the elements to be annihilated; (ii) to find simultaneously the $\lfloor \tfrac{1}{2}n \rfloor$ pairs of values for sine and cosine; (iii) to devise an alternative method of multiplying $J^{(k)} A^{(k)} J^{(k)\mathsf{T}}$ in order to take full advantage of the position of the nonzero elements in $J^{(k)}$. We consider each in turn.

(i)   For ease of programming, it is necessary to have a regular means of specifying the subsets Z_k to be annihilated at any given step. However, it is evident that a parallel implementation is not apparent in the above definition, primarily because the algebraic serial description disguises the 'pattern', i.e. the position at the kth and $(k + 1)$th steps of the elements to be annihilated.

Figure 6.1(a) shows the position of the elements for a matrix of order 16, and Fig. 6.1(b) shows how the position of the elements

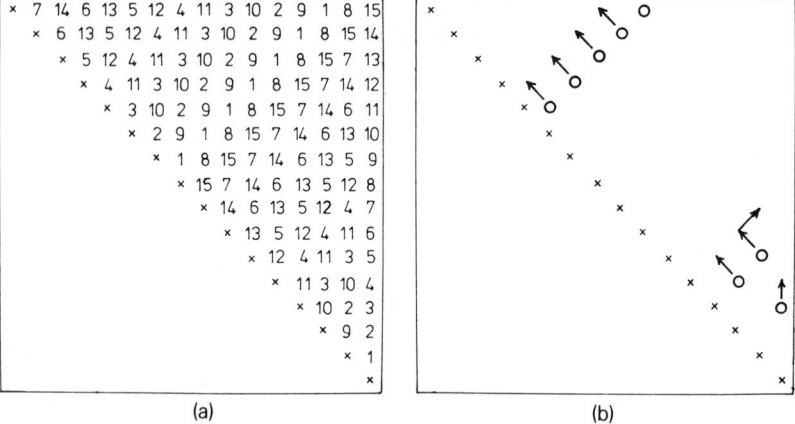

(a)   (b)

Fig. 6.1 Annihilation ordering and pattern generation for a 16×16 matrix

at the $(k+1)$th step can be deduced from that at the kth step. We indicate by o those elements annihilated at step 3 of the example in Fig. 6.1(a). The arrows indicate the movement required to generate the patterns for step 4 from those of step 3. (On parallel machines with the ability to perform rapid logical operations, this phase is relatively inexpensive.)

(ii)   Having found the location of the elements to be annihilated, the next step is to apply the similarity transformations as defined by (2.4). Care must be taken in the calculation of sine and cosine to avoid overflow and underflow. From (2.4) we have

$$\tan 2\alpha_{pq}^{(k)} = 2a_{pq}^{(k)}/(a_{pp}^{(k)} - a_{qq}^{(k)}).$$

One stable method of computing sine and cosine is as follows (Hammarling 1974; Rutishauser 1971; Parlett 1980):

$$\cot 2\alpha_{pq}^{(k)} = \tfrac{1}{2}(a_{pp}^{(k)} - a_{qq}^{(k)})/a_{pq}^{(k)}, \tag{6.1}$$

$$t_1 = 1/[|\cot 2\alpha_{pq}^{(k)}| + (1 + \cot^2 2\alpha_{pq}^{(k)})^{\frac{1}{2}}], \tag{6.2}$$

$$\cos \alpha_{pq}^{(k)} = 1/(1 + t_1^2), \qquad \sin \alpha_{pq}^{(k)} = \cos \alpha_{pq}^{(k)} t_1 \cdot \mathrm{sgn} \cot 2\alpha_{pq}^{(k)}. \tag{6.3}$$

For the parallel implementation we compute $\sin \alpha_{pq}^{(k)}$ and $\cos \alpha_{pq}^{(k)}$ for $\lfloor \tfrac{1}{2}n \rfloor$ pairs of elements simultaneously, using the above formulae.

Finally, we carry out the multiplication $J^{(k)}A^{(k)}J^{(k)\mathsf{T}}$, which involves two matrix multiplications each of $O(n^3)$ on a serial machine. By using the parallel matrix multiplier (see Chapter 2, Section 2.3), we can perform the multiplication in $O(n)$ steps, but the algorithm would then not take advantage of the position of the nonzero elements in $J^{(k)}$, which (as seen in Fig. 6.1) form a definite pattern. Any efficient matrix multiplication must exploit that pattern. We therefore propose such a scheme, which carries out both matrix multiplications in only six steps, provided that enough processors are available.

(iii)   *Schematic matrix multiplier*

At each iteration we are required to compute $A^{(k+1)} = J^{(k)}A^{(k)}J^{(k)\mathsf{T}}$. We shall consider this computation in two parts:

$$X^{(k+1)} = J^{(k)}A^{(k)} \quad \text{and} \quad A^{(k+1)} = X^{(k+1)}J^{(k)\mathsf{T}}.$$

(a)   *Computation of $X^{(k+1)}$*   If $J^{(k)}$ is the matrix defined in Section 5.4 with partition Z_k, we define a permutation α in cyclic notation as

$$\alpha(k) = \prod_{(p,q)\in Z_k} (p, q);$$

for example, from Fig. 6.1 with $k = 1$, we see that $\alpha(1) =$ $(1, 14)(2, 13)(3, 12) \ldots (16, 15)$. We introduce the notation for row and column permutation of a matrix M with α:

$$\overline{M}(i, j) = M(\alpha(i), j), \qquad \underline{M}(i, j) = M(i, \alpha(j)).$$

With this notation, the product $X^{(k+1)} = J^{(k)}A^{(k)}$ can be formed by termwise multiplication (.) as follows.

$$X^{(k+1)} = C^{(k)} \cdot A^{(k)} + S^{(k)} \cdot \overline{A}^{(k)},$$

where $C^{(k)}(i, j) = J^{(k)}(i, i)$ for all i and j, and

$$S^{(k)}(i, j) = \begin{cases} J^{(k)}(i, l) & \text{if } J^{(k)}(i, l) \neq 0 \text{ and } l \neq i, \\ 0 & \text{otherwise,} \end{cases}$$

for all i and j, being well defined since $\{l : J^{(k)}(i, l) \neq 0, l \neq i\}$ has one element at most. This algorithm takes three steps, and can be used to carry out the multiplication with the minimum arithmetic, but it also involves a permutation of the matrix $A^{(k)}$. Clearly, the multiplication scheme can be extended where there are more than two nonzero elements per column: this might be useful, for example, when tridiagonal or pentadiagonal matrices are being considered.

(b) *Computation of $A^{(k+1)}$* The product $X^{(k+1)}J^{(k)\mathsf{T}}$ can be computed in a similar manner as

$$A^{(k+1)} = C^{(k)\mathsf{T}} \cdot X^{(k+1)} + S^{(k)\mathsf{T}} \cdot \underline{X}^{(k+1)}.$$

In the next section, we discuss a new class of schemes referred to as *mobile schemes*. These lead to further simplification in organization, and highlight yet another approach which would probably apply to other problems in parallel processing.

5.7 Mobile schemes

A mobile scheme is simply a way of moving data around a multiprocessor array machine in a simple regular pattern—in the Jacobi context, it means incorporating the mapping of the rows and columns of the $n \times n$ matrix A into the hardware—so that, at any stage, the data required for a given calculation are near the appropriate processor. In this section, we present a general method to describe mobile schemes and two specific schemes for arbitrary n, the better of which reduces the relocation overhead to zero on a nearest-neighbour-connected architecture.

For Jacobi's method (and other plane-rotation algorithms), the essential problem is how to choose the rotation sequence and move the rows and columns so that the following requirements are met:

(a)   A sweep of $\frac{1}{2}n(n-1)$ rotations is completed in as few as possible parallel rotations ('stages'). Since at most $\lfloor \frac{1}{2}n \rfloor$ rotations can be performed in parallel, this will take at least $n-1$ stages if n is even, and n stages if n is odd.

(b)   Rows p and q of A are adjacent to each other whenever a rotation in the (p, q) plane is to be carried out. The rotation will be computed by the two rows of processors that hold the rows p and q of A. Here we assume that the system is fine-grained, that is, each element of the matrix A is allocated a processor. Clearly, a row or group of adjacent elements can be allocated a single processor (this is called coarse-granularity), and this may well suit some MIMD architectures.

(c)   Data movement between stages is short and uniform.

Conceptually, as explained in Section 5.6, each transformation stage falls into three parts. First we must extract, in parallel, all the triples $(\alpha_{p_i p_i}, \alpha_{q_i q_i}, \alpha_{p_i q_i})$ $(i = 1, \ldots, k_f)$, where k_f is the number of parallel rotations being carried out, and then compute, in parallel, the corresponding rotation coefficients c_i and s_i. Next, we perform the premultiplication of A by R, where $R^T = J^{(1)T} J^{(2)T} \ldots J^{(k_f)T}$, and finally the postmultiplication by R^T.

Mobile scheme MS1 is optimal as regards criteria (b) and (c) above and nearly optimal for (a). It is based on the 'caterpillar track', shown in Fig. 7.1 for the case $n = 9$. The numbers label

Column: 1   2   3   4   5   6   7   8   9

$$\cdot \leftarrow (2) \leftarrow \cdot \leftarrow (4) \leftarrow \cdot \leftarrow (6) \leftarrow \cdot \leftarrow (8) \leftarrow \cdot$$
$$\downarrow \qquad\qquad\qquad\qquad\qquad\qquad\qquad\qquad\qquad\qquad \uparrow$$
$$(1) \rightarrow \cdot \rightarrow (3) \rightarrow \cdot \rightarrow (5) \rightarrow \cdot \rightarrow (7) \rightarrow \cdot \rightarrow (9)$$

Fig. 7.1   Mobile scheme MS1

rows (and columns) of A. The column of Fig. 7.1 in which a number lies indicates which row of the hardware holds that row of A. At each stage, each number moves along to the next vacant space. Numbers that 'cross' define the plane rotations to be performed. For instance, the first stage moves the configuration to that shown in Fig. 7.2,

$$(2) \leftarrow \cdot \leftarrow (4) \leftarrow \cdot \leftarrow (6) \leftarrow \cdot \leftarrow (8) \leftarrow \cdot \leftarrow (9)$$
$$\downarrow \qquad\qquad\qquad\qquad\qquad\qquad\qquad\qquad\qquad \uparrow$$
$$\cdot \rightarrow (1) \rightarrow \cdot \rightarrow (3) \rightarrow \cdot \rightarrow (5) \rightarrow \cdot \rightarrow (7) \rightarrow \cdot$$

Fig. 7.2 Mobile scheme MS1 after the first stage

changing the row mapping as indicated:

$$\left|\begin{matrix} 1\ 2\ 3\ 4\ 5\ 6\ 7\ 8\ 9 \\ \downarrow\ 2\ 1\ 4\ 3\ 6\ 5\ 8\ 7\ 9. \end{matrix}\right.$$

Plane rotations are performed in the planes $(1,2)$, $(3,4)$, $(5,6)$, $(7,8)$. At the second stage, we rotate the configuration in Fig. 7.2 by one position anticlockwise to give the configuration shown in Fig. 7.3.

$$\cdot \leftarrow (4) \leftarrow \cdot \leftarrow (6) \leftarrow \cdot \leftarrow (8) \leftarrow \cdot \leftarrow (9) \leftarrow \cdot$$
$$\downarrow \qquad\qquad\qquad\qquad\qquad\qquad\qquad\qquad\qquad \uparrow$$
$$(2) \rightarrow \cdot \rightarrow (1) \rightarrow \cdot \rightarrow (3) \rightarrow \cdot \rightarrow (5) \rightarrow \cdot \rightarrow (7)$$

Fig. 7.3 Mobile scheme MS1 after the second stage

The row mapping is thereby changed according to

$$\left|\begin{matrix} 2\ 1\ 4\ 3\ 6\ 5\ 8\ 7\ 9 \\ \downarrow\ 2\ 4\ 1\ 6\ 3\ 8\ 5\ 9\ 7, \end{matrix}\right.$$

so that plane rotations are performed in the planes $(1,4)$, $(3,6)$, $(5,8)$, $(7,9)$. Similarly, at the third stage, we rotate the configuration in Fig. 7.3 by one further position anticlockwise to give the configuration shown in Fig. 7.4.

$$(4) \leftarrow \cdot \leftarrow (6) \leftarrow \cdot \leftarrow (8) \leftarrow \cdot \leftarrow (9) \leftarrow \cdot \leftarrow (7)$$
$$\downarrow \qquad\qquad\qquad\qquad\qquad\qquad\qquad\qquad\qquad \uparrow$$
$$\cdot \rightarrow (2) \rightarrow \cdot \rightarrow (1) \rightarrow \cdot \rightarrow (3) \rightarrow \cdot \rightarrow (5) \rightarrow \cdot$$

Fig. 7.4 Mobile scheme MS1 after the third stage

The row mapping change is

$$\begin{array}{l} \uparrow \quad 2\ 4\ 1\ 6\ 3\ 8\ 5\ 9\ 7 \\ \downarrow \quad 4\ 2\ 6\ 1\ 8\ 3\ 9\ 5\ 7, \end{array}$$

so that plane rotations are performed in the planes $(2, 4)$, $(1, 6)$, $(3, 8)$, $(5, 9)$. Continuing in this way, the fourth stage induces the row mapping change

$$\begin{array}{l} \uparrow \quad 4\ 2\ 6\ 1\ 8\ 3\ 9\ 5\ 7 \\ \downarrow \quad 4\ 6\ 2\ 8\ 1\ 9\ 3\ 7\ 5, \end{array}$$

so that plane rotations are performed in the planes $(2, 6)$, $(1, 8)$, $(3, 9)$, $(5, 7)$. Clearly, after precisely n stages, the rows have been exactly reversed in order, and each of the $\frac{1}{2}n(n-1)$ pairs (p, q) has crossed exactly once. The data movement on the hardware can be described with even greater simplicity: at each odd-numbered stage (counting from 1), we interchange and simultaneously rotate the rows of A currently in hardware rows 1 and 2, 3 and 4, and so on; at even-numbered stages, we interchange and rotate the contents of hardware rows 2 and 3, 4 and 5, and so on. The combination of exchange and rotation can be produced by using Givens matrices $\begin{bmatrix} s & c \\ c & -s \end{bmatrix}$ instead of rotation matrices $\begin{bmatrix} c & s \\ -s & c \end{bmatrix}$, so that the data movement overhead is reduced to zero by incorporating it into the arithmetic.

The only defect we see in MS1 is that (if n is even) it performs only $\frac{1}{2}n - 1$ rotations in parallel at even-numbered stages, instead of the maximum $\frac{1}{2}n$. This is not true of Brent and Luk's (1982) scheme, which is otherwise comparable to MS1, with the modification that row 1 is fixed and the remaining rows 2 to n move round a track at twice the speed of MS1; thus a new row is adjacent to row 1 at each stage, and 'rotates' with it.

Scheme MS2 also has simple and regular data movement. At odd-numbered stages we rotate in the planes $(1, n)$, $(2, n-1)$, and so on, and perform no data movement. At even-numbered stages we rotate in the planes $(2, n)$, $(3, n-1)$, and so on, and then carry out a cyclic shunt of the data, replacing the contents of

Table 7.1   Migration table for MS2 with $n = 8$

i	$k:$	1	2	3	4	5
1		1—	1	2—	2	
2		2-	2-	3-	3-	
3		3-	3-	4-	4-	
4		4-	4-	5-	5-	
5		5-	5	6-	6	*etc.*
6		6-	6-	7-	7-	
7		7-	7-	8-	8-	
8		8-	8-	1-	1-	

register (i, j) according to the scheme

$$(1, j) \leftarrow (j - 1, n), \qquad (i, j) \leftarrow (i - 1, j - 1) \quad (i = 2, ..., n),$$

for $j = 2, ..., n$.

The migration table for $n = 8$ is shown in Table 7.1. It is easy to see that MS2 is a sweep. There may be other data-movement schemes for implementing a Jacobi sweep, but it is hard to see how they could improve on MS1 and the Brent–Luk scheme for a lattice architecture.

The mobile schemes have particularly interesting sorting properties and can also be viewed as parallel sorting algorithms; these are discussed in Chapter 3, Section 3.8.

5.8 Reduced cyclic schemes

In this section we shall consider a slight modification to the parallel Jacobi method discussed earlier. Suppose we transform a real $n \times n$ symmetric matrix A° to the form

$$A = \begin{bmatrix} a_{11} & & & & & & a_{1n} \\ & a_{22} & & & & a_{2,n-1} & \\ & & a_{33} & & a_{3,n-2} & & \\ & & & \ddots & & & \\ & & a_{n-2,3} & & a_{n-2,n-2-} & & \\ & a_{n-1,2} & & & & a_{n-1,n-1-} & \\ a_{n1} & & & & & & a_{nn} \end{bmatrix}$$

through the application of Jacobi plane rotations. The orthogonal similarity transformations would therefore preserve the eigenvalues of A°, which may be obtained using the X form as follows: the characteristic equation $|A - \lambda I| = 0$ can be factorized to give

$$(a_{ii} - \lambda)(a_{n+1-i,n+1-i} - \lambda) - a_{i,n+1-i}^2 = 0 \quad (i = 1,\dots, \lfloor \tfrac{1}{2}n \rfloor). \quad (8.1)$$

If n is even (the case when n is odd is analogous) we obtain $\frac{1}{2}n$ quadratic equations, which must have real roots since the original matrix is real and symmetric.

We will call such a scheme a *reduced cyclic Jacobi method*, which can be summarized as follows:

Step 1 Perform the orthogonal similarity transformations

$$A^{(1)} = A^\circ, \qquad A^{(k+1)} = J^{(k)}A^{(k)}J^{(k)\mathsf{T}} \quad (k = 1,\dots, n - 2),$$

where $J^{(k)}$ is a compound rotation matrix (see Section 5.4) so that $\frac{1}{2}n$ plane rotations are performed simultaneously on the pivot elements whose locations are defined by Z_k, which does not include elements on the secondary diagonal, i.e. $a_{1,n}, a_{2,n-1},\dots, a_{n,1}$.

Repeat Step 1 until the general level of elements on the diagonal and the secondary diagonal is very much greater than that of the rest of the elements, according to (say) the tolerance condition

$$S \Big/ \left(\sum_{i,j=1}^{n} a_{ij}^2 - S \right) > \xi,$$

where $S = \sum_{i=1}^{n} (a_{ii}^2 + a_{i,n+1-i}^2)$; here $\xi = 10^p$, with p being chosen according to the requirement for accuracy.

Step 2 Obtain the eigenvalues of A° using eqn (8.1).

Each sweep consists of $n - 2$ steps compared with $n - 1$ for the optimal parallel Jacobi method. However, the reduction of one step necessitates the solving of $\frac{1}{2}n$ quadratic equations. Furthermore, we can see that, for Sameh's scheme (Fig. 6.1), the final subset of elements in the $(n - 1)$th step lies on the secondary diagonal. This was in fact the reason for choosing the X form: so that a parallel reduced cyclic Jacobi scheme could be developed by removing the $(n - 1)$th iteration. It is clear that we could have removed any one of the steps instead of the last. The method essentially reduces the eigenvalue problem for A° to a set of $\frac{1}{2}n$

disjoint 2×2 symmetric eigenvalue problems that are solved using the quadratic formula.

Naturally, we intend to determine whether one step of the parallel Jacobi method is expensive compared with the cost of solving $\frac{1}{2}n$ quadratic equations. But first we will discuss the design of the parallel reduced cyclic Jacobi method, which is almost identical to Sameh's scheme (see Section 5.6), with the following modifications:

(i)　Remove the $(n-1)$th iteration.
(ii)　Solve $\frac{1}{2}n$ quadratic equations simultaneously. We will take n to be even, but the scheme is analogous when n is odd.

By rearranging (8.1) as

$$\lambda^2 - (a_{ii} + a_{n+1-i,n+1-i})\lambda + a_{ii}a_{n+1-i,n+1-i} - a_{i,n+1-i}^2 = 0$$
$$(i = 1,\ldots, \tfrac{1}{2}n)$$

and letting $b_i = a_{ii} + a_{n+1-i,n+1-i}$ and $c_i = a_{ii}a_{n+1-i,n+1-i} - a_{i,n+1-i}^2$, the eigenvalues may be obtained as

$$\lambda_{i1} = \tfrac{1}{2}[b_i + (b_i^2 - 4c_i)^{\frac{1}{2}}], \qquad \lambda_{i2} = b_i - \lambda_{i1}. \tag{8.2}$$

We can simultaneously solve the quadratic equations as follows. Let

$$\boldsymbol{b} = \begin{bmatrix} a_{11} \\ a_{22} \\ \vdots \\ a_{nn} \end{bmatrix} + \begin{bmatrix} a_{nn} \\ a_{n-1,n-1} \\ \vdots \\ a_{11} \end{bmatrix},$$

$$\boldsymbol{c} = \begin{bmatrix} a_{11} \\ a_{22} \\ \vdots \\ a_{nn} \end{bmatrix} * \begin{bmatrix} a_{nn} \\ a_{n-1,n-1} \\ \vdots \\ a_{11} \end{bmatrix} - \begin{bmatrix} a_{1n} \\ a_{2,n-1} \\ \vdots \\ a_{n1} \end{bmatrix} * \begin{bmatrix} a_{1n} \\ a_{2,n-1} \\ \vdots \\ a_{n1} \end{bmatrix}$$

where $*$ indicates termwise multiplication. Equation (8.1) is then solved for all the roots, in parallel, using termwise operations (see Chapter 2, Section 2.1).

On the DAP, Sameh's original scheme took 1.602 sec for four sweeps in single-precision arithmetic. Therefore the best we could hope for is a reduction in time to $(1.602/63)\,62 = 1.577$ sec. Surprisingly, this is precisely the time taken using the reduced cyclic Jacobi method. Thus the extra cost of solving the $\frac{1}{2}n$

quadratic equations on the DAP is negligible in comparison with a single Jacobi iteration.

Finally we remark that, on a parallel machine, the reduced cyclic Jacobi method eliminates a single step per sweep of the parallel Jacobi method. On a serial machine, however, it eliminates $\frac{1}{2}n$ steps per sweep: a reduction in the total number of steps from $\frac{1}{2}n(n-1)$ to $\frac{1}{2}n(n-2)$.

5.9 Convergence properties

The roundoff error bounds for the parallel algorithm are somewhat better than for the general serial algorithm. It is shown in Modi (1982) that the computed eigenvalues are the exact eigenvalues of $A + E$ where

$$\|E\| \leq 2x(n-1)(1+x)^{2n-4} \|A\|,$$

where $x = 6 \cdot 2^{-t}$ and t is the machine precision. The factor $n-1$ is replaced by $n^{\frac{3}{2}}$ in the corresponding analysis by Wilkinson (1965).

In this section, we outline the current position as far as convergence is concerned for the following variants of the Jacobi method: (i) the classical Jacobi method, (ii) special cyclic Jacobi schemes, (iii) the general cyclic Jacobi method, (iv) a restricted-angle general cyclic Jacobi method with bound ϕ, where $\phi \in (0, \frac{1}{4}\pi)$, (v) the modified general cyclic Jacobi method, and (vi) the parallel Jacobi scheme incorporating the threshold strategy.

(i) *The classical Jacobi method*

In the classical Jacobi method, the element $a_{pq}^{(k)}$ $(p < q)$ to be annihilated is chosen so that it corresponds to the maximum off-diagonal element of A, i.e.

$$|a_{pq}^{(k)}| = \max_{i \neq j} |a_{ij}^{(k)}|. \tag{9.1}$$

The angle θ_k is chosen so that

$$\tan 2\theta_k = \frac{2a_{pq}^{(k)}}{a_{pp}^{(k)} - a_{qq}^{(k)}}, \qquad -\tfrac{1}{4}\pi < \theta_k \leq \tfrac{1}{4}\pi; \tag{9.2}$$

θ_k is uniquely defined unless $a_{pq}^{(k)} = a_{pp}^{(k)} - a_{qq}^{(k)} = 0$, when it can be taken to be zero. It can be shown that, if (9.1)–(9.2) are satisfied, then $A^{(k)}$ converges $(k \to \infty)$ to a diagonal matrix representing the eigenvalues of A. The ultimate convergence is quadratic (Henrici 1958).

(ii) *Special cyclic Jacobi schemes*

In the case of special cyclic Jacobi schemes, the pairs (p_i, q_i) are defined according to the schemes given in Section 5.3. For both row and column cyclic ordering, θ_k is chosen to satisfy (9.2). In each case, the ultimate rate of convergence is quadratic (Forsythe and Henrici 1960; Wilkinson 1962; Van Kempton 1966).

(iii) *The general cyclic Jacobi method*

The general cyclic Jacobi method selects the pairs (p, q) in *any cyclic order*, with θ_k satisfying (9.2). In other words, if

(i)   the pairs (p, q) are chosen in any cyclic order,
(ii) $\tan 2\theta_k = 2a_{pq}^{(k)}/(a_{pp}^{(k)} - a_{qq}^{(k)})$ $(p < q)$,
(iii) $-\tfrac{1}{4}\pi < \theta_k \leqslant \tfrac{1}{4}\pi$,

then we call such a scheme a *general cyclic Jacobi method*.

The proof of convergence of a general cyclic Jacobi method has never been established. If (9.2) is satisfied, and $\theta_k = 0$ in the degenerate case, then it is not known whether satisfactory convergence will theoretically always occur. However, Brodlie and Powell (1975) do give an example of failure to converge when, in the degenerate case, we are free to make any choice of θ_k in $(-\tfrac{1}{4}\pi, \tfrac{1}{4}\pi]$.

(iv) *A restricted-angle general cyclic Jacobi method with bound ϕ in $(0, \tfrac{1}{4}\pi)$*

A restricted-angle general cyclic Jacobi method with bound ϕ $(0 < \phi < \tfrac{1}{4}\pi)$ is one in which the pairs (p, q) are chosen in any cyclic order, and the angle θ_k is chosen as follows:

(i)   calculate $\bar{\theta}_k$ as for θ_k in (9.2);
(ii) if $|\bar{\theta}_k| > \phi$, then $\theta_k = (\operatorname{sgn} \bar{\theta}_k)\phi$; otherwise $\theta_k = \bar{\theta}_k$.

Henrici (1958) proved that such a method will converge if ϕ satisfies

$$\sin \phi < 1/(n-2)(n^2 - n - 1) \quad (n \geqslant 3). \tag{9.3}$$

Brodlie and Powell (1975) have shown that convergence of the restricted-angle general cyclic method takes place when the angle restriction is less severe than (9.3):

$$\sin \phi < \frac{(n^2 - 2n - 1)^{\frac{1}{2}} - (n - 3)}{2(n - 2)} \quad (n \geqslant 3). \tag{9.4}$$

From (9.3) and (9.4) we note that Henrici's bound is $O(1/n^3)$ and Brodlie and Powell's is $O(1/n)$.

Although convergence is guaranteed by either (9.3) or (9.4), the imposition of an angle restriction prevents immediate annihilation of large pivot elements, and thus decreases the rate of convergence considerably. Brodlie and Powell have suggested a slight modification which applies the angle restriction only when $|a_{pq}^{(k)}|$ and $|a_{pp}^{(k)} - a_{qq}^{(k)}|$ are small. This preserves the usual efficiency of the ordinary Jacobi method.

(v)   *The modified general cyclic Jacobi method*
Let η be a small positive number (e.g. 10^{-4} times the modulus of the maximum off-diagonal element). Choose ϕ as in (9.4) (in practice we may take $\phi = 1/(n - 1)$ $(n \geqslant 3)$). Then:

(a)   if $|a_{pq}^{(k)}| \geqslant \eta$, define θ_k by (9.2);
(b)   otherwise, define $\bar{\theta}_k$ as for θ_k in (9.2) and set

$$\theta_k = \begin{cases} (\text{sgn } \bar{\theta}_k)\phi & \text{if } |\bar{\theta}_k| > \phi, \\ \bar{\theta}_k & \text{otherwise} \end{cases}$$

(thus, the angle restriction is imposed only when the two diagonal elements are close and the off-diagonal element is small in modulus).

(vi)   *The parallel Jacobi scheme incorporating the threshold strategy*

Theorem 9.1   For a parallel Jacobi scheme where the noninteracting plane rotations $R_1, \ldots, R_m$ $(m \in \mathbb{Z})$ are applied simultaneously on the pivots $a_{p_r, q_r}^{(k)}$ $(r = 1, \ldots, m)$ of a real symmetric matrix $A^{(k)}$ of order $n > 2$, it is sufficient to guarantee convergence if one of the $a_{p_r, q_r}^{(k)2}$ exceeds $cs(A^{(k)})$, where c is a positive constant and

$$s(A^{(k)}) = \frac{1}{2} \sum_{i \neq j} a_{ij}^{(k)2} \quad (k = 1, 2, \ldots).$$

Proof. For noninteracting rotations with pivots $(p_1, q_1), \ldots,$ (p_m, q_m), we have

$$s(A^{(k+1)}) = s(R_{p_m q_m} \cdots R_{p_1 q_1} A^{(k)} R_{p_1 q_1}^{\mathsf{T}} \cdots R_{p_m q_m}^{\mathsf{T}}) = s(A^{(k)}) - \sum_{r=1}^{m} a_{p_r q_r}^{(k)2}$$

$$= m^{(k)} s(A^{(k)}), \quad (9.5)$$

where

$$m^{(k)} = 1 - \sum_{r=1}^{m} a_{p_r q_r}^{(k)2} / s(A^{(k)}).$$

Observe that, if one of the $a_{p_r q_r}^{(k)2}$ exceeds $cs(A^{(k)})$, then $m^{(k)} < 1 - c$. Now iterate (9.5):

$$s(A^{(k+1)}) = m^{(k)} m^{(k-1)} m^{(k-2)} \cdots m^{(1)} s(A^{(1)}) = \left(\prod_{i=1}^{k} m^{(i)} \right) s(A^{(1)}).$$

Then, since $\prod_{i=1}^{k} m^{(i)} < (1 - c)^k$, we obtain

$$s(A^{(k+1)}) \rightarrow 0 \quad \text{as} \quad k \rightarrow \infty.$$

Since the sum of the squares of the off-diagonal elements converges to zero, the matrix $A^{(k)}$ converges to a fixed diagonal matrix D as $k \rightarrow \infty$.

5.10 Approximate annihilation

The use of a mobile scheme, discussed in Section 5.7, minimizes organizational costs. The cost of the parallel Jacobi algorithm is therefore dominated by the cost of the arithmetic operations in the main loop, which consist of the computation of the coefficients (c, s) in the plane rotations and the two matrix multiplications. The relative cost of computing (c, s) is significantly greater on a parallel computer. In comparison, the cost of broadcasting is very small. Table 10.1 shows the major cost components for both serial and parallel computers. The fourth column shows the fractional cost of computing (c, s) on both architectures. In the case of parallel machines, this ratio is independent of n (the order of the matrix) since the number of operations taken for multiplication does not depend on the matrix size, provided that enough processors are available. Timing measurements on the DAP indicate that this cost can be up to half the total arithmetic cost. It is for this reason that, in

Table 10.1   Relative arithmetic costs for the Jacobi method on serial and parallel machines

	Total arithmetic cost per iteration		
	Multiplication (1)	Computation(c, s) (2)	$(2)/[(1) + (2)]$
Serial:	$O(n)$	$O(1)$	$O(n^{-1})$
Parallel:	$O(1)$	$O(1)$	$\frac{1}{2}$ (in practice)

computing (c, s), we consider the use of approximate formulae, which are cheaper to compute than exact formulae, so that worthwhile speed-up is obtained. Some of the early work on using simpler formulae to compute (c, s) on serial computers was done by Wilkinson (1965).

If $A = \begin{bmatrix} a & b \\ b & d \end{bmatrix}$, $L = \begin{bmatrix} c & -s \\ s & c \end{bmatrix}$, $c = \cos \theta$, and $s = \sin \theta$, then

$$L^{\mathsf{T}} A L = \begin{bmatrix} a' & b' \\ b' & d' \end{bmatrix},$$

where

$$b' = \frac{(d - a)t}{1 + t^2} + \frac{b(1 - t^2)}{1 + t^2}$$

and $t = \tan \theta$. Note that $(a^2 + 2b^2 + d^2)^{\frac{1}{2}}$ is left unchanged by this transformation. A measure of the suitability of a scheme for approximately annihilating b is an upper bound k for

$$|b'|/|b| \tag{10.1}$$

which is given in the theorems of this section.

The exact formula (making b' vanish) is $t = \tau$, where τ is chosen as the smaller root, in absolute value, of $t/(1 - t^2) = b/(a - d)$. Call the right-hand side σ; then $\tau = \tau(\sigma)$ and $\tau^2 \sigma + \tau - \sigma = 0$. Thus

$$\tau = 2\sigma/[(1 + 4\sigma^2)^{\frac{1}{2}} + 1].$$

We note that

$$\tau(\sigma)/\sigma \to 1 \quad \text{as} \quad \sigma \to 0. \tag{10.2}$$

If this convergence is retained in the approximate formula for τ, then, in practice, convergence of the algorithm will be assured. In order to make the formula reasonable over the entire range, it is preferable that it should also be exact at infinity:

$$\tau(\sigma) \to \pm 1 \quad \text{as} \quad \sigma \to \pm\infty. \tag{10.3}$$

We now list three examples of an approximate $\tau(\sigma)$ satisfying (10.2).

Formula 1 Let

$$\tau = \tau_1(\sigma) = \sigma/(1 + |\sigma|),$$
$$= [b/(|a - d| + |b|)] \, \text{sgn}(a - d),$$

with, as always,

$$c = 1/(1 + \tau^2)^{\frac{1}{2}}, \qquad s = \tau/(1 + \tau^2)^{\frac{1}{2}} = c\tau.$$

Here (10.3) is satisfied as well as (10.2): the bound k for (10.1) is indicated in Theorem 10.1.

Formula 2 Let

$$s = \sigma(1 + \alpha |\sigma|)/(1 + \gamma |\sigma| + \delta\sigma^2),$$
$$c = (1 + \beta |\sigma| + \alpha\sigma^2)/(1 + \gamma |\sigma| + \delta\sigma^2),$$

so that

$$\tau = \tau_2(\sigma) = \sigma(1 + \alpha |\sigma|)/(1 + \beta |\sigma| + \alpha\sigma^2),$$

where α, β, γ, δ are chosen so that $c^2 + s^2 = 1$; it turns out that

$$\beta = \gamma = 2\alpha = \sqrt{2}\delta = \sqrt{2} + 1.$$

Again, both (10.2) and (10.3) are satisfied, and no square-root calculations are involved. But there are more terms and therefore greater cost, and the bound k in (10.1) is worse than for Formula 1. We shall therefore not discuss this approximation further.

Formula 3 Let

$$\tau = \tau_3(\sigma) = \sigma,$$

with, as before,

$$c = 1/(1 + \tau^2)^{\frac{1}{2}}, \qquad s = \tau/(1 + \tau^2)^{\frac{1}{2}} = c\tau.$$

Here (10.3) is not satisfied, but we can use this approximation if we are in the range of convergence, as explained in Theorem 10.2.

Theorem 10.1 If the approximate formula $\tau = \sigma/(1 + |\sigma|)$, where $\sigma = b/(a - d)$, is used instead of the exact formula in order to reduce b, then

$$|b'| \leq k |b|,$$

where $k = \frac{1}{2}(\sqrt{2} - 1)$ and b' is the value of b after the transformation.

Proof. We have

$$b'/b = (\sigma - \tau - \sigma\tau^2)/\sigma(1 + \tau^2),$$
$$= |\sigma|/(1 + 2 |\sigma| + 2\sigma^2).$$

It is easily seen that the maximum of this occurs when $\sigma = \frac{1}{2}\sqrt{2}$, giving $b'/b = \frac{1}{2}(\sqrt{2} - 1)$ as asserted.   □

Figure 10.1 illustrates the measure of error for Formula 1.

Theorem 10.2 To ensure that the transformation $b \mapsto b'$ satisfies the bound

$$|b'| < |b|/n,$$

the application of the approximate formula $\tau = \sigma$ in the Jacobi algorithm should be restricted to points where

$$|b|/|a - c| < 1/m, \qquad m^2 + 1 = n.$$

Proof. We have

$$b' = [(d - a)\tau + b(1 - \tau^2)]/(1 + \tau^2),$$

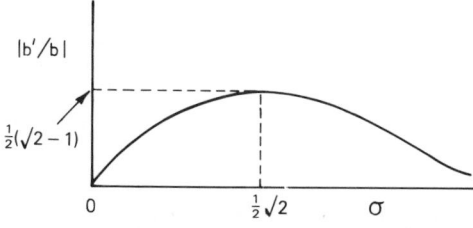

Fig. 10.1 Error propagation using Formula 1

where $d - a = -b/\sigma$ and $\tau = \sigma$; so $b'/b = -\sigma^2/(1 + \sigma^2)$. To ensure that $|b'| < |b|/n$, we have $|\sigma| < 1/m$, where $m^2 = n - 1$, which gives

$$|b|/|a - d| < 1/m,$$

as asserted.   □

We note that Formula 1 gives $b'/b = O(\sigma)$, whereas Formula 3 gives $b'/b = O(\sigma^2)$, so that the error incurred by Formula 3 is smaller. However, as indicated in Theorem 10.2, we can only use Formula 3 at the point where $|\sigma| < 1/m$. In the implementation, we choose m to be 3 so that the maximum error incurred is ten per cent.

The measure $|b'/b|$ in (10.1) bears no relation to whether or not the process has converged. It simply provides a measure of how much the off-diagonal element b is reduced.

The ratio $|\sigma'/\sigma|$, where $\sigma = b/(a - d)$, measures the convergence of the Jacobi process. It takes account of the difference in eigenvalues due to the approximate annihilation. Simple algebra shows that

(i)   for Formula 1, $|\sigma'/\sigma| < 0.15$⎫

(ii) for Formula 3, $|\sigma'/\sigma| < 0.25$⎭'   (10.4)

which indicates rapid convergence.

Timing

Table 10.2 gives the operation count on the DAP using the exact and the approximate formulae.

Table 10.2   Operation count and timings for exact and approximate formulae

	Computation of (c, s)	Time (msec)
Exact:	2 $\uparrow^2$, 2∗, 2 $\sqrt{}$, 3÷, 4±	1.488
Formula 1:	1 $\uparrow^2$, 1∗, 1 $\sqrt{}$, 2÷, 3±	0.904
Formula 3:	1 $\uparrow^2$, 1∗, 1 $\sqrt{}$, 2÷, 2±	0.849

Figure 10.2 shows the rate of decrease of the sum of squares of the off-diagonal elements after each sweep, for the 64×64

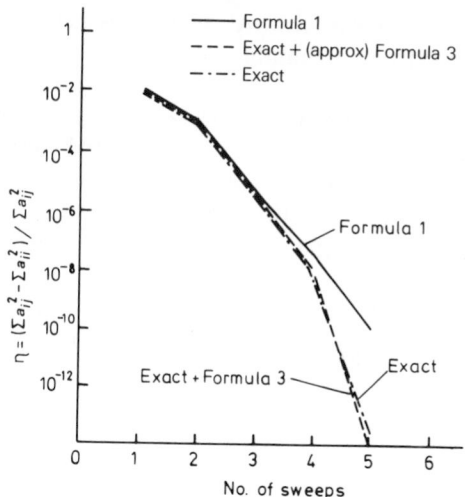

Fig. 10.2 Comparison of rate of decrease of off-diagonal elements using various formulae

matrix of elements

$$a_{ij} = \begin{cases} i^2 + 64 & (i = j), \\ i + j & (i \neq j). \end{cases}$$

The measure $\eta = (\sum a_{ij}^2 - \sum a_{ii}^2)/\sum a_{ij}^2$ is used. It is seen that, for the combination of the exact formula and (the approximate) Formula 3, the value of η approximates very closely that of the exact formula.

In conclusion, if the point of application is chosen carefully so that the error incurred by its use is zero to machine precision, then there is no loss of accuracy. Indeed, no extra iterations are needed, and there is an overall saving in computational time. Formula 3 has the added advantage over Formula 1 that it can differentiate between nearly equal eigenvalues.

5.11 A direct method for completing eigenproblem solutions

Here we propose replacement of the final sweep(s) by a single application of the perturbation method which affects all off-diagonal elements simultaneously; in the limit, this is as effective as an entire sweep of the Jacobi method. The method is valid when the matrix eigenvalues are distinct, which holds for the matrices arising from a number of engineering problems of practical importance.

Let A be a close-to-diagonal $n \times n$ real symmetric matrix with small off-diagonal elements and distinct diagonal elements, which are close to the corresponding eigenvalues (also distinct). We obtain an approximate solution to the eigenvalue problem by constructing a matrix U and a diagonal matrix Δ such that

$$UU^T \approx I, \qquad UA \approx \Delta U. \qquad (11.1\text{a,b})$$

Write $A = A_0 + A_1$, where A_0 is diagonal and A_1 has zeros on its main diagonal. Solve, for the skew-symmetric matrix V, the equation

$$A_0 V - V A_0 = A_1, \qquad (11.2)$$

so that

$$v_{ii} = 0, \qquad v_{ij} = \frac{a_{ij}}{a_{ii} - a_{jj}} \quad (i \neq j). \qquad (11.3)$$

Now write

$$\tfrac{1}{2}(VA_1 - A_1 V) = B = B_0 + B_1, \qquad (11.4)$$

where B_0 is diagonal and B_1 has zero diagonal; these matrices are symmetric. Solve, for the skew-symmetric matrix W, the equation

$$A_0 W - W A_0 = B_1, \qquad (11.5)$$

so that

$$w_{ii} = 0, \qquad w_{ij} = \frac{b_{ij}}{a_{ii} - a_{jj}} \quad (i \neq j). \qquad (11.6)$$

Our original version of the proposed approximations was

$$U = I + V + W + \tfrac{1}{2}V^2, \qquad \Delta = A_0 + B_0. \qquad (11.7\text{a,b})$$

We note that (11.2) gives

$$VA_1 + A_1 V = A_0 V^2 - V^2 A_0. \qquad (11.8)$$

The results of some experimentation (see below) suggest that there is improved accuracy if (11.7a) is replaced by

$$U = e^X = I + X + \tfrac{1}{2}X^2 + \tfrac{1}{6}X^3 + \tfrac{1}{24}X^4 + \cdots, \qquad (11.9)$$

where $X = V + W$; this will be unitary to machine precision if sufficient terms are taken. We may also replace (11.7b) by

$$\Delta = UAU^T. \tag{11.10}$$

The use of (11.9) and (11.10) is recommended in computational practice. An alternative possibility is, for example,

$$U = (I + \tfrac{1}{2}X)(I - \tfrac{1}{2}X)^{-1}$$
$$= I + X + \tfrac{1}{2}X^2 + \tfrac{1}{4}X^3 + \tfrac{1}{8}X^4 + \cdots, \tag{11.11}$$

which is also unitary. It is not clear why these are significantly better than (11.7a), nor whether one among such series as (11.9) and (11.11) is generally preferable; these all start with $I + X + \tfrac{1}{2}X^2$.

When A is not close to diagonal, the technique is to apply the Jacobi method up to the point at which the matrix becomes diagonally dominant, and then make one application of the method described above; the resulting eigenvector matrix will thus be UR^T, where R is the product of Jacobi rotations which transform A to a sufficient degree of diagonal dominance, as indicated below. If many terms of $X, X^2, X^3, \ldots$ are taken, it will be worse than using just one term and repeating the process, since the convergence of the Jacobi method is quadratic and $\varepsilon^{2^n} < (k\varepsilon)^n$ for some n $(0 < k, 0 < \varepsilon < 1)$.

General estimates and justifications of (11.1)

Let

$$\max\{|a_{ij}| : i \neq j\} = \alpha, \qquad \min\{|a_{ii} - a_{jj}| : i \neq j\} = \delta.$$

Writing $m(X)$ for $\max_{i,j} |x_{ij}|$, from (11.3), (11.4), (11.6) we see that

$$m(V) \leq \alpha/\delta, \qquad m(B) \leq n\alpha^2/\delta, \qquad m(W) \leq n\alpha^2/\delta^2. \tag{11.12}$$

Other estimates will be derived by using the inequality (for $n \times n$ matrices)

$$m(XY) \leq nm(X)m(Y). \tag{11.13}$$

Now suppose that

$$\max\{n^3\alpha^4/\delta^4, n^2\alpha^3/\delta^2, n^2\alpha^4/\delta^3\} = \varepsilon \ll 1. \tag{11.14}$$

From (11.7a) and the skew symmetry of V and W, it follows that

$$UU^T = (I + V + W + \tfrac{1}{2}V^2)(I - V - W + \tfrac{1}{2}V^2) = I + E,$$

where

$$E = \tfrac{1}{4}V^4 + \tfrac{1}{2}(WV^2 - V^2W) - (VW + WV),$$

and then (11.12)–(11.14) give

$$m(E) \leqslant \tfrac{1}{4}n^3\alpha^4/\delta^4 + n^3\alpha^4/\delta^4 + 2n^2\alpha^3/\delta^3 \leqslant \tfrac{13}{4}\varepsilon.$$

From the equations defining U and Δ, together with the identity (11.8), it follows by straightforward matrix algebra that

$$UA = U\Delta + F,$$

where

$$F = WA_1 + \tfrac{1}{2}A_1V^2 - B_0V - \tfrac{1}{2}V^2B_0,$$

and then (11.12)–(11.14) yield

$$m(F) \leqslant n^2\alpha^3/\delta^3 + \tfrac{1}{2}n^2\alpha^3/\delta^2 + n^2\alpha^3/\delta^2 + \tfrac{1}{2}n^2\alpha^4/\delta^3 \leqslant 3\varepsilon. \quad (11.15)$$

It also follows that UAU^T is close to Δ, so that its off-diagonal elements are small. In particular, these will certainly all be smaller than α (so that UAU^T is closer to diagonal, in one sense, than A) if $\varepsilon \ll \alpha$. One test for this would be that $\alpha \ll \min\{\delta^{\frac{3}{4}}/n, \delta/n\}$. In practice, the direct method seems remarkably accurate for rather larger values of α. We have chosen to take $\varepsilon > 10^{-3}$ as a sensible test for applicability of the direct method, for the matrices considered.

The formulae (11.7a,b) may easily be arrived at by applying the perturbation method in the following form. Assume that $U = I + X_1 + X_2 + Z$, where the elements of X_1, X_2, and Z are respectively of orders 1, 2, and $\geqslant 3$ in the off-diagonal elements of A. If this matrix U is to be approximately unitary: that is, to satisfy (11.1a), then substitution in (11.1a) and equating to zero terms of orders 1 and 2 gives

$$X_1 + X_1^T = 0, \qquad (X_2 + X_2^T) - X_1^T X_1 = 0.$$

From the first equation, it follows that X_1 is a skew-symmetric matrix V, and from the second, that $X_2 - \tfrac{1}{2}V^2$ is a skew-symmetric matrix W. Thus we expect to use a formula $U =$

$I + V + W + \frac{1}{2}V^2$, where V and W are skew-symmetric matrices of orders 1 and 2 in the off-diagonal elements of A. Substitution in the equation $UAU^T \approx \Delta$, where Δ is an unknown diagonal matrix, then leads to the formulae (11.2) and (11.5) for V and W, and also yields the value (11.7b) for Δ.

It is not clear to what extent the condition $m(F) \leqslant 3\varepsilon$ of (11.5) could yield accurate theoretical estimates of the exact position of the eigenvalues of A. It follows from Gerschgorin's theorems (Wilkinson, 1965, pp. 71–2) that the Gerschgorin intervals, with centres a_{ii} and radii $\sum_{j \neq i} |a_{ij}|$, are disjoint provided that $\delta \geqslant 2(n-1)\alpha$. Since $n^3 \alpha^4 / \delta^4 \leqslant \varepsilon$, this condition will already have been satisfied at the stage when our 'direct method' is applied, provided that $\varepsilon \leqslant n^3 / 16(n-1)^4$. This is true for the examples considered below, where $n = 64$ and $\varepsilon = 10^{-3}$.

Algorithm

The perturbation method discussed above may be combined with the Jacobi process as follows. Assuming $[a_{ij}]$ to be an $n \times n$ symmetric matrix with distinct eigenvalues, first apply the standard parallel Jacobi process (see Section 5.4). At the end of each sweep, test whether the condition for applying the direct method is satisfied, as follows:

compute $\alpha = \max\{|a_{ij}| : i \neq j\}$ and $\delta = \min\{|a_{ii} - a_{jj}| : i \neq j\}$;
compute $\varepsilon = \max\{n^3 \alpha^4 / \delta^4, n^2 \alpha^3 / \delta^2, n^2 \alpha^4 / \delta^3\}$.

If $\varepsilon \geqslant 10^{-k}$ (where $k = 3$, see above), continue with a further Jacobi sweep; otherwise, apply the direct method, as follows.

express $A = A_0 + A_1$, where A_0 is diagonal and A_1 has zero main diagonal,
compute V, where $v_{ii} = 0$ and $v_{ij} = a_{ij}/(a_{ii} - a_{jj})$ $(i \neq j)$;
compute $B = \frac{1}{2}(VA_1 - A_1 V)$;
compute W, where $w_{ii} = 0$ and $w_{ij} = b_{ij}/(a_{ii} - a_{jj})$ $(i \neq j)$;
compute $X = V + W$;
compute $U = e^X = I + X + \frac{1}{2}X^2 + \frac{1}{6}X^3 + \cdots$;
compute diag UAU^T, with elements the eigenvalues;
compute $U^T R^T$, with columns the eigenvectors.

We note that U^T need not be calculated explicitly, and that R^T is the product of plane rotations obtained using the Jacobi method.

Implementation

We report on tests carried out on two 64×64 real symmetric matrices:

(1) $[a_{ij}]$ with $a_{ij} = i + j$ $(i \neq j)$ and $a_{ii} = i^2 + 64$, which is diagonally dominant and is known to have distinct eigenvalues in the range $(0.2, 6778)$;

(2) $[a_{ij}]$ with $a_{ij} = 0$ $(|i - j| \neq 1)$ and $a_{i,i+1} = a_{i+1,i} = 1$, which is tridiagonal and has eigenvalues in the range $(-2, 2)$, some of which are very close.

Denoting by J_n the application of n sweeps of the Jacobi method, it was found that J_6 or J_7 was sufficient to guarantee 16 significant figures, using double precision (64 bits), providing a benchmark to test the results of taking fewer sweeps, followed by a single application of the direct method. For the relevant values of n ($= 64$), α, and δ, we have $\varepsilon = n^2\alpha^3/\delta^2$, and an appropriate test for whether it might be advantageous to discontinue Jacobi sweeps and apply the direct method is that ε should be sufficiently small, as indicated earlier.

The use of $\alpha = \max\{|a_{ij}| : i \neq j\}$ and $\delta = \min\{|a_{ii} - a_{jj}| : i \neq j\}$ renders such a test quite unsuitable for serial computation because of the high cost of computing the extrema of a set of real numbers. This is not true for a bit-oriented machine such as the DAP, because the algorithm for extrema depends on the bit length of the numbers and not on the size of the set, provided that enough processing units are available, as is the case here. (To find which, among a set of numbers given in binary form, are maximal, successively eliminate as candidates those having 0 in the highest binary place, if any have 1; then those having 0 in the next place down, if any retained candidates have 1; and so on.)

In brief, the computation of δ is carried out as follows:

(i)   extract the vector $\boldsymbol{d} = (a_{11}, \ldots, a_{nn})$ (one parallel step);

(ii) form two matrices, one with all rows equal to $\boldsymbol{d}^{\mathrm{T}}$ and the other with all columns equal to $\boldsymbol{d}$ (each only one step, using the row and column 'highways'), and take their difference (one step);

(iii) replace the zero diagonal elements by a large constant (one step);

(iv) find the minimum of the absolute values of the elements of this matrix.

The computation of α is similar.

On the DAP, the calculations for V and W ((11.3) and (11.6)) are inexpensive in terms of time, using techniques like (i) and (ii) above. The matrix multiplications ((11.4) and (11.7a)) incur the major cost, although the parallel algorithm used takes only $O(n)$ steps instead of the $O(n^3)$ for serial methods. If the eigenvalues (Δ) alone are required, and not the eigenvectors, then (11.7b) gives Δ without the necessity of calculating W. Only one matrix multiplication VA_1 is required (since $-A_1V$ is the transpose), whereas V^2 is needed to compute U, and altogether it takes three multiplications to compute both the eigenvalues and the matrix of eigenvectors $U^T R^T$. Each matrix multiplication takes less than 37 μsec, so the direct method takes under 37 μsec for the eigenvalues only, and under 0.11 sec for the complete eigensolution—less than the 0.2 sec required for one complete Jacobi sweep.

We first examined the matrix defined by $a_{ij} = i + j$ $(i \neq j)$, with $a_{ii} = i^2 + 64$, and found that the 64 eigenvalues can be obtained to 16 significant figures by J_6 (six sweeps of the Jacobi method), and are (in increasing order) approximately

$$e_1 = 0.20, \quad e_2 = 63.51, \quad e_3 = 65.94, \quad \dots \,,$$
$$e_{63} = 3997.15, \quad e_{64} = 6777.10.$$

Convergence turned out to be slowest for the first three. In Table 11.1, we show (as $J_p + \Delta$) the effect of applying the direct method in the form (11.7b) after p ($= 4$ or 5) Jacobi sweeps.

The reason why convergence remains slow, after four sweeps, for the small eigenvalues is likely to be that the neighbourhood surrounding them has relatively large elements. Figure 11.1 shows which elements of A are greater than 10^{-3} after four sweeps, with 1 indicating elements less than 10^{-3}.

The large eigenvalues (lower right corner) have already converged. After five sweeps, all off-diagonal elements are less than 10^{-3} and, as seen from Table 11.1, the condition for applying the direct method ($\varepsilon \ll 1$) is now satisfied, yielding the expected improvement in the eigenvalues.

In Table 11.2 (and Table 11.3 below), $J_p + \Delta_q$ denotes the result when p Jacobi sweeps are followed by an application of the direct method in the form (11.10), $\Delta = UAU^T$, where U is

Table 11.1 Application of the direct method, using (11.7b), with $\alpha = \max\{|a_{ij}| : i \neq j\}$, $\delta = \min\{|a_{ii} - a_{jj}| : i \neq j\}$, and $\varepsilon = n^2 \alpha^3 / \delta^2$

Method	α	δ	ε	e_1	e_2	e_3
J_4	0.3445	2.4298	28.3787	0.20448 43844 847891	63.51082 52617 6269	65.94058 17412 7515
$J_4 + \Delta$				0.20106 46047 892698	63.51043 17037 6667	65.93980 96669 8373
J_5	0.0005	2.4294	0.7 E−7	0.20105 89751 991050	63.51042 43161 1070	65.93981 23918 2791
$J_5 + \Delta$				0.20105 89720 322694	63.51042 42201 2616	65.93981 24717 4584
J_6	0.6 E−8	2.4294	0.16 E−21	0.20105 89720 322660	63.51042 42201 2458	65.93981 24717 4818

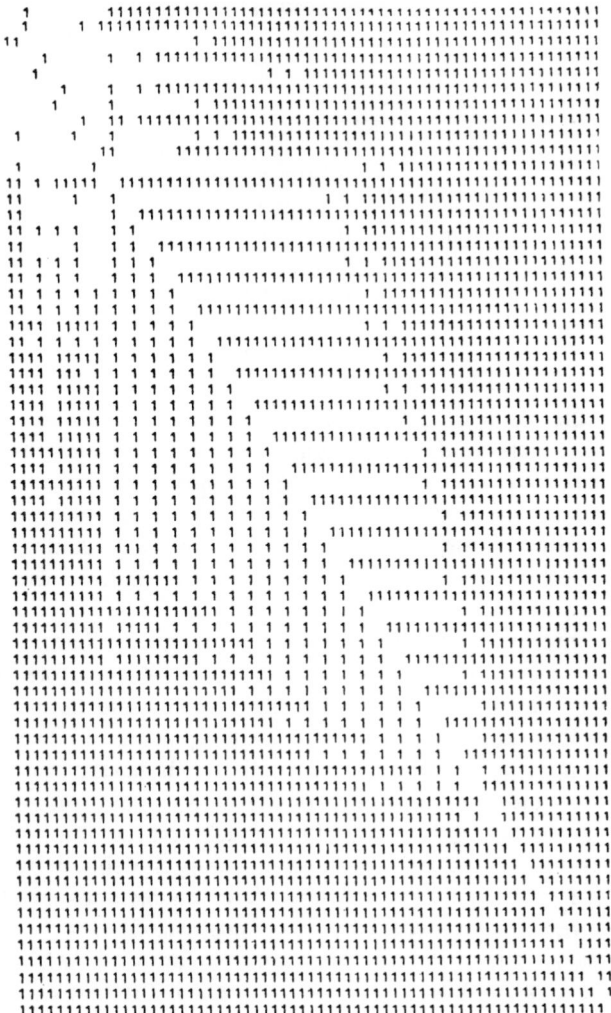

Fig. 11.1 Elements with $|a_{ij}| \geqslant 10^{-3}$ after four sweeps

calculated by taking the terms up to X^q in the series (11.9) for e^X, where $X = V + W$. We indicate results for e_1, but similar results were obtained for other eigenvalues.

Comparing Tables 11.1 and 11.2, we see that there is a distinct improvement in accuracy, particularly from J_4 to $J_4 + \Delta_2$.

Table 11.2   Application of the direct
method to e_1

Method	e_1
$J_4 + \Delta$	0.20106 46047 892698
$J_4 + \Delta_2$	0.20105 89726 989845
J_5	0.20105 89751 991050
$J_5 + \Delta$	0.20105 89720 322694
$J_5 + \Delta_2$	0.20105 89720 322661
J_6	0.20105 89720 322660

Indeed, the result is significantly better than $J_4 + \Delta$, and $J_5 + \Delta_2$ gives the exact values. In general, therefore, it seems that it is worth making U unitary, enabling us to apply the direct method earlier (than with nonunitary U). $J_5 + \Delta_2$ gives the exact results, but its computation incurs slightly more cost than $J_5 + \Delta$. (Similar results were obtained by making U unitary, using the formulae (11.11).)

It is interesting to note that $J_4 + \Delta_2$ gives slightly more accurate results than J_5, and that we could therefore apply $(J_4 + \Delta_2) + \Delta_2$, which would give exact results and take slightly less time than $J_5 + \Delta_2$. This would entail replacing two sweeps by repeated application of the direct method. However, in general, the application of Δ_q may corrupt the data, unless q is so large that U, given by (11.9), is unitary to machine precision.

We examined the 64×64 matrix $A = [a_{ij}]$ defined by $a_{ij} = 0$ ($|i - j| \neq 1$) and $a_{i,i+1} = a_{i+1,i} = 1$. The eigenvalues of A are given (in decreasing order) by $\lambda_k = 2 \cos(k\pi/65)$ ($k = 1,\ldots, 64$); numerically, their values are approximately 1.998, 1.991, 1.979,..., 0.048, -0.048,..., -1.998, of which the first (and last) few are relatively close. They can be obtained to 16 significant figures by J_7. After five sweeps, all the off-diagonal elements were less than 10^{-3}, and the effects of applying the direct method were calculated (although ε was still approximately 0.03). Table 11.3 indicates results for the closest eigenvalues λ_1 and λ_2, but similar results were obtained for the others.

It is seen that the method was effective in spite of the relative closeness of some eigenvalues ($\delta = 0.003$, while after J_5 the

Table 11.3 Application of direct method to λ_1 and λ_2

Method	λ_1	λ_2
J_5	1.99766 41802 45964	1.99066 32569 63184
J_6	1.99766 44536 64534	1.99066 32694 35234
$J_5 + \Delta$	1.99766 44536 25066	1.99066 32694 79780
$J_5 + \Delta_3$	1.99766 44536 63848	1.99066 32694 34497
$J_5 + \Delta_4$	1.99766 44536 64585	1.99066 32694 35237
J_7	1.99766 44536 64579	1.99066 32694 35233

value of α was approximately 0.0007), but more terms were needed in (11.9) in order to make U effectively unitary.

5.12 The generalized eigenvalue problem

In this section, we consider the problem of computing the eigensolutions (λ, z) of the generalized problem $Az = \lambda Mz$, $z \neq 0$, where A and M are $n \times n$ matrices. This problem arises in many important physical areas such as dynamical analysis and electrical networks. In the former case, A is the stiffness matrix, M is the mass matrix, z represents the shape of vibration, and $\lambda^{\frac{1}{2}}$ the frequency of vibration. We will discuss various techniques for computing the eigensolutions, having first introduced some terminology.

(i) λ is an *eigenvalue of the pair* (A, M), that is, a solution of $\det(A - \lambda M) = 0$; z is an *eigenvector*.

(ii)   Two pairs (A, M) and $(\bar{A}, \bar{M})$ are *equivalent* if there exist nonsingular matrices E and F such that

$$\bar{A} = EAF, \qquad \bar{M} = EMF.$$

Since

$$\det(EAF - \lambda EMF) = (\det E)(\det F) \det(A - \lambda M),$$

with $\det E$ and $\det F$ nonzero, the characteristic polynomials of the two problems are multiples of each other, and the eigenvalues are the same. The eigenvectors are simply related: if (λ, y) is an eigensolution of $EAFy = \lambda EMFy$, then $A(Fy) = \lambda M(Fy)$, and so $y = F^{-1}z'$, where z' is an eigenvector, corresponding to the eigenvalue λ, of the pair (A, M).

The theory behind the generalized problem is much more difficult than in the case of the standard problem, and we shall not develop it here except to mention some of its unusual features:

(a) The characteristic equation $\det(A - \lambda M) = 0$ may vanish identically, for example, when A and M have a common null vector. If $A = M = \begin{bmatrix} a & 0 \\ 0 & 0 \end{bmatrix}$, then $(1, x)$ and (ξ, e_2) are eigensolutions, where the eigenvalue ξ corresponding to e_2 can take any value, and the eigenvector x can be any nonzero vector.

(b) All coefficients, except the constant term, of the characteristic polynomial may vanish, giving us an 'infinite' eigenvalue corresponding to a well-defined eigenvector.

(c) For symmetric A and M, we can obtain complex eigensolutions.

We now consider some numerical techniques for solving the generalized problem with the restriction that M or some combination $\alpha A + \beta M$ $(\beta \neq 0)$ is positive definite so that the above three phenomena do not arise. Both A and M are assumed to be symmetric.

(i) Reduction to the standard eigenvalue problem

Recent research has generally been directed towards reducing the generalized problem to an equivalent standard problem, when highly efficient methods become available:

(a) Form $M^{-1}A$ (with the possible loss of symmetry and sparsity).

(b) Solve the standard eigenvalue problem, via $M = JD^2J^T$, where J is an orthogonal matrix and D is a diagonal matrix. We have

$$A - \lambda M = JD(D^{-1}J^TAJD^{-1} - \lambda I)DJ^T,$$

that is, $A - \lambda M = JD(\bar{A} - \lambda I)DJ^T$, where $\bar{A} = D^{-1}J^TAJD^{-1}$, so that the generalized problem reduces to the standard form $\bar{A}x = \lambda x$, where $x = DJ^Tz$.

(c) Form the Choleski decomposition $M = LL^T$, with L

nonsingular lower triangular, so that

$$A - \lambda M = L(\hat{A} - \lambda I)L^\mathsf{T},$$

where $\hat{A} = L^{-1}AL^{-\mathsf{T}}$, and $\hat{A}y = \lambda y$, where $y = L^\mathsf{T}z$.

(ii) *Simultaneous reduction to diagonal form*

As an extension of the orthogonal similarity transformation used to solve the standard eigenvalue problem (Section 5.2), it is natural to seek an analogous form for the pair (A, M). The result, aptly termed *simultaneous reduction to diagonal form*, states that for certain pairs (A, M) there exists an invertible matrix R such that

$$R^\mathsf{T}AR = D = \mathrm{diag}(d_1, \dots, d_n), \qquad R^\mathsf{T}MR = S = \mathrm{diag}(s_1, \dots, s_n).$$

Although the ratios d_i/s_i $(i = 1, \dots, n)$ are unique (up to a permutation), the matrices D and S are not necessarily so. A method based on this result is discussed below.

(iii) *Polynomial and vector iteration*

From the polynomial $p(\lambda) = \det(A - \lambda M) = 0$, the zeros of $p(\lambda)$ are the eigenvalues, and the eigenvectors can be obtained using *inverse iteration*, which computes eigenvectors for given approximate eigenvalues. Indeed the *power method* can also be used to find eigenvectors corresponding to the few largest eigenvalues. For the case of tridiagonal A and M, the two-term linear recurrence relation can be evaluated in parallel using recursive doubling; see Chapter 2, Section 2.5.

We now proceed to discuss in detail one technique in group (ii) (for detailed description of the others the reader is referred to Golub and Van Loan 1983; Parlett 1980; Stewart 1973; and Wilkinson 1965).

The method we consider is essentially an extension of the standard parallel Jacobi method for real symmetric A and M with M or some linear combination $\alpha A + \beta M$ $(\beta \neq 0)$ positive definite. It avoids reduction to the standard problem and is particularly useful when A and M are ill conditioned or the off-diagonal elements are small. The serial method uses a transformation in the (i, j)-plane to annihilate the (i, j)-elements of A and M. In the parallel method, a group of similarity transformations is simultaneously applied to annihilate $\frac{1}{2}n$ elements of A and M at each step. This is repeated for different sets of $\frac{1}{2}n$ pivot elements; after five or six complete sweeps, both A and M converge to diagonal matrices, from which the eigenvalues appear as the quotients $\lambda_i = a_{ii}^{(s)}/m_{ii}^{(s)}$ $(i = 1, \dots, n)$, where s is the number of sweeps, while the eigenvectors are the columns of the product of the transformation matrices.

Parallel Jacobi algorithm for the generalized problem

The algorithm is a simple extension of the standard Jacobi method. The serial method is discussed in Bathé and Wilson (1976). We describe it briefly and fix our notation.

Let $i \neq j$ and let T be the $n \times n$ matrix with 1's in the diagonal, $T_{ij} = -\alpha$ and $T_{ji} = \beta$, and all other elements 0. In order to annihilate the (i, j)-elements of both TAT^{T} and TMT^{T}, we must have

$$\alpha = \delta_i/v, \qquad \beta = \delta_j/v,$$

where v satisfies the quadratic equation

$$v^2 - \delta_{ij}v - \delta_i\delta_j = 0,$$

and where

$$\delta_i = \begin{vmatrix} a_{ii} & a_{ij} \\ m_{ii} & m_{ij} \end{vmatrix}, \qquad \delta_j = \begin{vmatrix} a_{jj} & a_{ij} \\ m_{jj} & m_{ij} \end{vmatrix}, \qquad \delta_{ij} = \begin{vmatrix} a_{ii} & a_{jj} \\ m_{ii} & m_{jj} \end{vmatrix}.$$

If M is positive definite and if δ_{ij} and $\delta_i\delta_j$ are not both zero, then the solutions for v are real and at least one is nonzero, and $\det T = 1 + \alpha\beta \neq 0$. In order to keep α and β small we choose the root v of largest magnitude.

Such transformations are applied in succession to A and M until they both converge to diagonal form. Thus we obtain

$$NAN^{\mathsf{T}} = \operatorname*{diag}_{1 \leq i \leq n} a_{ii}^{(s)}, \qquad NMN^{\mathsf{T}} = \operatorname*{diag}_{1 \leq i \leq n} m_{ii}^{(s)},$$

where $N = T_s \cdots T_1$, so that

$$N(A - \lambda M)N^{\mathsf{T}} = \operatorname*{diag}_{1 \leq i \leq n} (a_{ii}^{(s)} - \lambda m_{ii}^{(s)}).$$

Looking at the jth column of the equation

$$(A - \lambda M)N^{\mathsf{T}} = N^{-1} \operatorname{diag} (a_{ii}^{(s)} - \lambda m_{ii}^{(s)})$$

for $\lambda = a_{jj}^{(s)}/m_{jj}^{(s)}$, we see that $(n_{j1}, \ldots, n_{jn})^{\mathsf{T}}$ is an eigenvector corresponding to this eigenvalue.

In the parallel method, a sweep is divided into $n-1$ blocks consisting each of $\frac{1}{2}n$ disjoint transformations, and no transformation is repeated within a sweep. Instead of a single transformation with matrix T, here $\frac{1}{2}n$ independent transformations having a product of matrices equal to F (say) are performed simultaneously. (We may assume that n is even.)

The parallel generalized Jacobi method thus proceeds as follows.

Let $Z = \{(p, q) : 1 \leq p < q \leq n\}$ be partitioned into subsets $Z_1, \ldots, Z_{n-1}$, each of size $\frac{1}{2}n$, such that, if (p, q) and (p', q') are distinct pairs in Z_k, then $\{p, q\} \cap \{p', q'\} = \emptyset$. A single sweep of the method is defined by:

$$A^{(1)} = A \quad \text{and} \quad M^{(1)} = M.$$

For $k = 1, \ldots, n-1$, calculate

$$A^{(k+1)} = F^{(k)}A^{(k)}F^{(k)\mathsf{T}}, \qquad M^{(k+1)} = F^{(k)}M^{(k)}F^{(k)\mathsf{T}}.$$

After a sweep is completed, the cycle is repeated with the initial A and M replaced by the latest $A^{(k+1)}$ and $M^{(k+1)}$ computed in the previous sweep, until both A and M converge to the diagonal form to the required degree of accuracy (see *tolerance* level, defined below). Here,

$$F^{(k)}(p, q) = -\alpha^{(k)}(p, q) \quad \text{and} \quad F^{(k)}(q, p) = \beta^{(k)}(p, q),$$

for $(p, q) \in Z_k$, and

$$F^{(k)}(i, j) = 0 \ (i \neq j) \quad \text{and} \quad F^{(k)}(i, i) = 1, \quad \text{for } (i, j) \notin Z_k;$$

where

$$\alpha^k(p, q) = \delta_p^{(k)}/v^{(k)}, \qquad \beta^{(k)}(p, q) = \delta_q^{(k)}/v^{(k)},$$

with

$$\delta_p^{(k)} = a_{pp}^{(k)}m_{pq}^{(k)} - a_{pq}^{(k)}m_{pp}^{(k)}, \qquad \delta_q^{(k)} = a_{qq}^{(k)}m_{pq}^{(k)} - a_{pq}^{(k)}m_{qq}^{(k)},$$

$$\delta_{pq}^{(k)} = a_{pp}^{(k)}m_{qq}^{(k)} - a_{qq}^{(k)}m_{pp}^{(k)},$$

and with

$$v^{(k)} = \begin{cases} \frac{1}{2}[\delta_{pq}^{(k)} + (\operatorname{sgn} \delta_{pq}^{(k)})(\delta_{pq}^{(k)2} + 4\delta_p^{(k)}\delta_q^{(k)})^{\frac{1}{2}}] & \text{if } \delta_{pq}^{(k)} \neq 0, \\ (\delta_p^{(k)}\delta_q^{(k)})^{\frac{1}{2}} & \text{if } \delta_{pq}^{(k)} = 0; \end{cases}$$

in the trivial case when also $\delta_p^{(k)} \delta_q^{(k)} = 0$, we may take

$$\beta^{(k)}(p, q) = 0, \qquad \alpha^{(k)}(p, q) = m_{pq}^{(k)}/m_{qq}^{(k)}.$$

Convergence test

Convergence is measured as follows.

(i) Compare successive eigenvalue approximations at each step:

$$|\lambda_i^{(t+1)} - \lambda_i^{(t)}|/|\lambda_i^{(t+1)}| \leq 10^{-s} \quad (i = 1,\dots, n),$$

where s is set to required accuracy, and

$$\lambda_i^{(t)} = a_{ii}^{(t)}/m_{ii}^{(t)}, \qquad \lambda_i^{(t+1)} = a_{ii}^{(t+1)}/m_{ii}^{(t+1)}.$$

(ii) Test if off-diagonal elements are small: set

$$\rho_1 = \sum_{\substack{i,j \\ i \neq j}} a_{ij}^2 \Big/ \sum_{i,j} a_{ij}^2, \qquad \rho_2 = \sum_{\substack{i,j \\ i \neq j}} m_{ij}^2 \Big/ \sum_{i,j} m_{ij}^2,$$

and check if ρ_1 and ρ_2 are of the order of machine precision.

Conditions (i) and (ii) are tested after each sweep; if both are satisfied, then convergence is assumed.

Numerical properties

At present, convergence analysis of the Jacobi method for the generalized problem appears to have been restricted to the special cyclic orderings (see Section 5.3). Zimmerman (1969) has shown in this case that, if convergence occurs, it is ultimately quadratic. This does not appear to have been proved, however, for a general cyclic ordering. When M is the identity matrix, the method reduces to the standard Jacobi method; in this case, quadratic convergence has been established for the threshold scheme and the angle-restriction method (see Section 5.9).

By using the parallel method, the transformation to the standard problem is avoided. This is particularly desirable if the off-diagonal elements of A and M are small, when the annihilation of elements will not change the diagonal elements greatly, nor the ratio $\lambda_i = a_{ii}^{(s)}/m_{ij}^{(s)}$ $(i = 1,\dots, n)$. Furthermore, fast convergence can be expected if the matrix is diagonally dominant, and this is why the generalized Jacobi method has proved effective in *subspace iteration*.

The details of implementation on a mesh-connected architecture and the application of approximate annihilation schemes are discussed by Modi *et al.* (1984).

6 QR factorization

6.1 Introduction

This short chapter is a preface to the following one on singular-value decomposition and least squares, where QR factorization forms the core of the computation. QR factorization has recently received a lot of attention, particularly in the development of Givens sequences that are suited for parallel computation. After the introductory remarks, we consider some of these new orderings, and discuss the ease and effectiveness with which they can be implemented. Finally we compare the relative efficiency of Givens sequences (using plane rotations) and Householder reflections for achieving the same factorization.

For $m \geqslant n$, an $m \times n$ matrix A can be factorized by applying a Givens sequence, i.e. any sequence of Givens rotations in which zeros, once created, are preserved: $A = QR$ where Q is an $m \times m$ orthogonal matrix, $R = \begin{bmatrix} \tilde{U} \\ 0 \end{bmatrix}$ is an $m \times n$ matrix, and $\tilde{U}$ is an $n \times n$ upper triangular matrix.

The standard Givens process for determining R involves calculating $Q^{\mathsf{T}} A$ by choosing Q^{T} to be a product of a sequence of plane rotations each of which annihilates an element of A below the main diagonal without destroying the previously introduced zeros. In eliminating the (i, j)-element of A, the only rows affected are i and $i - 1$, and we premultiply these by the matrix $P = \begin{bmatrix} c & s \\ -s & c \end{bmatrix}$, where $c = \cos \theta$ and $s = \sin \theta$:

$$\begin{bmatrix} c & s \\ -s & c \end{bmatrix} \begin{bmatrix} a_{i-1,1} & a_{i-1,2} & \cdots & a_{i-1,j} & \cdots & a_{i-1,n} \\ a_{i,1} & a_{i,2} & \cdots & a_{i,j} & \cdots & a_{i,n} \end{bmatrix}.$$

For the updated $a_{i,j}$ to be zero, we require $-sa_{i-1,j} + ca_{i,j} = 0$, from which we obtain

$$s = a_{i,j}/(a_{i,j}^2 + a_{i-1,j}^2)^{\frac{1}{2}}, \qquad c = a_{i-1,j}/(a_{i,j}^2 + a_{i-1,j}^2)^{\frac{1}{2}};$$

the elements in rows $i-1$ and i are updated accordingly.

The sequential process of triangularization consists of the following steps (Givens 1958).

For $r = 1,\ldots, n$, compute

$$Q_r = \prod_{i=0}^{m-r-1} Q_{m-i,r}, \qquad A_r = \left(\prod_{i=0}^{r-1} Q_{r-i}\right)A,$$

where $Q_{p,r}$ $(p = r+1,\ldots, m)$ is the Givens rotation performed between the rows p and $p-1$ so as to annihilate the (p,r)-element of

$$\left(\prod_{i=1}^{p-r-1} Q_{p-i,r}\right)A_r$$

(an empty product is taken as the identity). Then, setting

$$R = A_n, \qquad Q^{\mathsf{T}} = Q_n \cdots Q_1,$$

we obtain the factorization $A = QR$.

For further details see Wilkinson (1965) and Golub and Van Loan (1983).

6.2 Givens sequences

An equivalent version of the Givens sequence for performing QR factorization of an $m \times n$ matrix A, with $m \geq n$, can be defined as follows.

List the set $S = \{(p, q) : 1 \leq q < p \leq m, q \leq n\}$ as

$$\left((p_k, q_k) : k = 1,\ldots, mn - \tfrac{1}{2}n^2 - \tfrac{1}{2}n\right)$$

in the order

$$(m, 1),(m-1, 1),\ldots, (2, 1); (m, 2),(m-1, 2),\ldots, (3, 2);$$

$$\ldots; \ldots; (m, n), (m-1, n),\ldots, (n+1, n).$$

The pair (p_k, q_k) indicates that a Givens rotation $G(p_k, q_k)$ is performed between the rows p_k and $p_k - 1$ so as to annihilate the (p_k, q_k) element of $G(p_{k-1}, q_{k-1}) \cdots G(p_1, q_1)A$.

Gentleman's (1975) method (the *standard* ordering) partitions the set S into subsets S_r $(r = 1,..., m + n - 2)$ in the order

$$\{(m, 1)\} \quad \{(m - 1, 1)\} \quad \{(m, 2), (m - 2, 1)\}$$
$$\{(m - 1, 2), (m - 3, 1)\} \quad \{(m, 3), (m - 2, 2), (m - 4, 1)\},...,$$

(2.1)

so that

$$S_r = \{(p, q) : 1 \leq q < p \leq m, q \leq n, m + 2q = p + r + 1\}.$$

The rotations in S_r are disjoint, and therefore they commute and can be performed simultaneously. The corresponding product matrix is

$$Q^{(r)} = \prod \{G(p, q) : (p, q) \in S_r\},$$

where now $G(p, q)$ is defined according to the pair order in (2.1). For example, for $m \geq 5$ and $n \geq 5$, we have

$$Q^{(5)} = G(m, 3)G(m - 2, 2)G(m - 4, 1).$$

The sequence (2.1) for a 16×5 matrix is illustrated in Fig. 2.1(i). An integer r is entered where zeros are created at the rth step.

6.3 The new scheme

The new scheme is best explained by reference to a particular example: taking the same 16×5 matrix, we indicate in Fig. 2.1(ii) the stages at which elements are annihilated.

At the first stage, we are able to annihilate simultaneously the eight elements in positions

$$(16, 1), (15, 1),..., (9, 1) \tag{3.1}$$

by rotating between the rows $16, 15,..., 9$ and the rows $8,..., 1$ in a given order, because these rotations are evidently disjoint.

At the second stage, we annihilate simultaneously the elements in positions

$$(8, 1), (7, 1), (6, 1), (5, 1)$$

and also

$$(16, 2), (15, 2), (14, 2), (13, 2),$$

by rotations between the rows 8, 7, 6, 5 and 4, 3, 2, 1 for the first

15				
14	16			
13	15	17		
12	14	16	18	
11	13	15	17	19
10	12	14	16	18
9	11	13	15	17
8	10	12	14	16
7	9	11	13	15
6	8	10	12	14
5	7	9	11	13
4	6	8	10	12
3	5	7	9	11
2	4	6	8	10
1	3	5	7	9

(i)

4				
3	6			
3	5	8		
2	5	7	10	
2	4	7	9	12
2	4	6	9	11
2	4	6	8	10
1	3	5	8	10
1	3	5	7	9
1	3	5	7	9
1	3	4	6	8
1	2	4	6	8
1	2	4	5	7
1	2	3	5	7
1	2	3	4	6

(ii)

Fig. 2.1 (i) Standard Givens ordering and (ii) new scheme applied to a 16×5 matrix

four, and between the rows 16, 15, 14, 13 and 12, 11, 10, 9 for the second four. Not only are these rotations evidently disjoint, but also they preserve the already created zeros in position (3.1). The principle is that, once a zero has been created in a certain position (p, q), it is preserved thereafter, because any subsequent rotation involving row p is between this row and another, say p', such that the element in position (p', q) is already zero.

At the third stage, we are able to annihilate the elements in positions

$$(4, 1), (3, 1); \quad (12, 2), (11, 2), (10, 2), (9, 2); \quad (16, 3), (15, 3);$$

and so on. Only twelve stages are required to complete the process, as compared to 19 for the standard scheme.

We now specify the new scheme in terms of $Z(i, r)$, the number of zeros in column i after r steps.

Define $Z(i, r)$ $(i = 1,..., n; r = 0, 1,...)$ by

$$Z(i, 0) = 0 \quad (i = 1,..., n), \qquad Z(0, r) = m \quad (r = 0, 1,...),$$

$$Z(i, r + 1) = Z(i, r) + \lfloor \tfrac{1}{2}Z(i - 1, r) - \tfrac{1}{2}Z(i, r) \rfloor \quad (i = 1,..., n),$$
$$(3.2)$$

where $\lfloor \ \rfloor$ denotes integer part.

After completion of the rth step, we already have $Z(i, r)$ zeros in the last $Z(i, r)$ places of the ith column, and $Z(i - 1, r)$ zeros in the last $Z(i - 1, r)$ places of the $(i - 1)$th column. We also have $Z(1, r) \geqslant Z(2, r) \geqslant \cdots \geqslant Z(n, r)$.

Let A_i denote the set of rows in which, after completion of the rth step, a zero has already been created in column $i - 1$ but not yet in column i. Then $|A_i| = Z(i - 1, r) - Z(i, r)$.

Let the set B_i consist of the last $\lfloor \tfrac{1}{2}|A_i| \rfloor$ rows in A_i, and C_i the preceding $\lfloor \tfrac{1}{2}|A_i| \rfloor$. At step $r + 1$, we can introduce zeros into all the positions (b, i) with $b \in B_i$, by rotating each row from B_i against a row from C_i, and this can be done simultaneously for all values of i for which $Z(i - 1, r) - Z(i, r) > 1$.

Analysis

Let $p_{i,j}$ denote the number of annihilated elements in column j after i parallel steps. Then the relation (3.2) gives the recurrence

$$p_{i,j} = p_{i-1,j} + \lfloor \tfrac{1}{2}(p_{i-1,j-1} - p_{i-1,j}) \rfloor \quad (i = 1, 2,...),$$

where $p_{i,0} = m$, for all i, and $p_{i,j} = 0$ $(j = 1,..., n)$.

This is equivalent to $p_{i,j} = \lfloor \tfrac{1}{2}(p_{i-1,j-1} + p_{i-1,j}) \rfloor$, and so $p_{i,j} \leqslant mq_{i,j}$, where $q_{i,j} = \tfrac{1}{2}(q_{i-1,j-1} + q_{i-1,j})$ with $q_{i,0} = 1$. Writing $Q(x, y) = \sum q_{i,j} x^i x^j$ for the generating function we find

$$Q(x, y) = (1 - \tfrac{1}{2}x)(1 - x)^{-1}[1 - \tfrac{1}{2}x(1 + y)]^{-1},$$

and expansion gives

$$q_{i,j} = 1 - 2^{-i} \sum_{k=0}^{j-1} \binom{i}{k}.$$

Let r be the number of parallel rotations required to produce $m - n$ zeros in column n. Then a lower bound on r is given by the solution of

$$mq_{r-1,n} < m - n, \qquad mq_{r,n} \geq m - n,$$

i.e. $r = \lfloor t \rfloor$, where t is the solution of $2^{-t} \sum_{k=0}^{n-1} \binom{t}{k} = \dfrac{n}{m}$.

From the identity

$$2^{-t} \sum_{k=s}^{t} \binom{t}{k} = \binom{t}{s} 2^{-(t+1)} \left(1 + \frac{t+1}{s+1} 2^{-1} + \frac{(t+1)(t+2)}{(s+1)(s+2)} 2^{-2} + \cdots \right),$$

it is straightforward to show that

$$1 - q_{t,n} = 2^{-t} \binom{t}{n-1} \left(1 + \frac{\theta(n-1)}{t} + O(t^{-2})\right), \qquad 0 < \theta < 1.$$

To find the number of parallel rotations to triangularize an $m \times n$ array, we need to solve, for t, the equation $1 - q_{t,n} = n/m$. (The difference between $p_{i,j}$ and $q_{i,j}$ can be shown to be negligible as $m \to \infty$.) Suppose m and hence t are large compared to n. Then

$$\binom{t}{n-1} \left(1 + \frac{n-1}{t}\right) \approx \frac{t^{n-1}}{(n-1)!},$$

and so, asymptotically, $2^{-t} t^{n-1} = n!/m$ is to be solved for t.

Taking logarithms to base 2, we get

$$-t + (n-1)\log_2 t = \log_2 n! - \log_2 m.$$

If n is fixed, then

$$t \sim \log_2 m + (n-1)\log_2 t \quad (m \to \infty),$$

so a first approximation is $t \approx \log_2 m$ and, taking the second approximation, we have

$$t \sim \log_2 m + (n-1)\log_2 \log_2 m \quad (m \to \infty).$$

From the above approximate analysis, it is clear that this scheme takes fewer stages than the standard Givens scheme. The number of stages required for various values of m and n is listed in Table 3.1. Taking a 100×6 example, the new scheme takes 20 stages whereas the standard scheme takes $100 + 6 - 2 = 104$. The scheme shows the greatest improvement for $m \gg n$.

Table 3.1   Number of stages required for QR factorization of an $m \times n$ matrix A $(m > n)$ using the new scheme

m	$n:$ 2	4	6	8	10	12	14	16
5	4	6	—	—	—	—	—	—
10	6	10	12	14	—	—	—	—
25	8	12	16	20	24	28	32	35
50	9	14	18	22	26	30	34	38
75	10	15	19	23	27	31	35	39
100	10	15	20	24	28	32	36	40

6.4   Fibonacci schemes

For $i = 1, 2, \ldots$, let $(u_k^i : k = -i + 1, -i + 2, \ldots)$ denote the sequence determined by the initial conditions

$$u_0^i = \cdots = u_{-i+1}^i = 0 \tag{4.1}$$

and the recurrence relation

$$u_k^i = u_{k-1}^i + \cdots + u_{k-i}^i + 1 \quad (k = 1, 2, \ldots). \tag{4.2}$$

Our interest will be in the terms $u_1^i, u_2^i, \ldots$, and some of these are shown in Fig. 4.1 together with the sums $s_k^i = u_1^i + \cdots + u_k^i$.

Consider the following infinite triangular array of integers A^i. At the top of the first column is a single 0; below that are $u_2^i = 2$ successive copies of -1; below these, u_3^i copies of -2; and so on. The second column is like the first except that all entries are increased by $i + 1$, and the whole column moved down one place. In an exactly similar way, the third column is derived from the second, and so on. The array A^2 is illustrated in Fig. 4.2.

$u_k^1 = k$	1	2	3	4	5	6	7	8
s_k^1	1	3	6	10	15	21	28	36
u_k^2	1	2	4	7	12	20	33	54
s_k^2	1	3	7	14	26	46	79	133
u_k^3	1	2	4	8	15	28	52	96
s_k^3	1	3	7	15	30	58	110	206
u_k^4	1	2	4	8	16	31	60	116
s_k^4	1	3	7	15	31	62	122	238

Fig. 4.1   Fibonacci schemes

```
 0    .   .   .    .    .
-1    3   .   .    .    .
-1    2   6   .    .    .
-2    2   5   9    .    .
-2    1   5   8   12    .
-2    1   4   8   11   15
-2    1   4   7   11   14 | 18
-3    1   4   7   10   14 | 17  21
-3    0   4   7   10   13 | 17  20
-3    0   3   7   10   13 | 16  20
-3    0   3   6   10   13 | 16  19
-3    0   3   6    9   13 | 16  19
-3    0   3   6    9   12 | 16  19
-3    0   3   6    9   12 | 15  19
-4    0   3   6    9   12 | 15  18

-4   -1   3   6    9   12   15  18
```

Fig. 4.2 Annihilation ordering for Fibonacci scheme A^2

To apply the Givens scheme derived from A^i to an $m \times n$ matrix, place an $m \times n$ grid on A^i, with the left-hand top corner over the position immediately above the 0 in the corner of A^i (the case $m = 16$ with $n = 6$ is indicated by the dotted lines in Fig. 4.2). If the integer in the left-hand bottom corner is u, say, then add $u + 1$ to all the integers in the grid. It is easy to see that what is obtained is the diagram (analogous to Fig. 2.1(ii)) of a Givens scheme, which we call F^i.

Empirically, the schemes F^1, F^2,... are found to be less efficient than that of Section 6.3. However, they are formed in a very systematic and easily specified way, and are susceptible of rather exact analysis because of the simplicity of the relation (4.2). The connection with Fibonacci numbers for $i = 2$ suggested the proposed name.

If $f^i(m, n)$ denotes the number of steps taken using F^i for an $m \times n$ matrix, then

$$f^i(m, n + 1) = f^i(m, 1) + (i + 1)(n - 1),$$

and it is thus sufficient to estimate the numbers $f^i(m, 1)$.

Putting $i = 1$ in (4.2) gives $u_k^1 = k$ for $k = 1, 2,...$; thus $s_k^1 = \frac{1}{2}k(k + 1)$ and so $f^1(m, 1) \approx (2m)^{\frac{1}{2}}$. Hence

$$f^1(m, n) \approx (2m)^{\frac{1}{2}} + 2n$$

(better than Gentleman's $m + n - 2$ for large $m > n$).

For $i > 1$, the usual methods of solution for (4.1)–(4.2) give a formula for u_k^i with leading term $a\alpha(i)^k$, where a is a constant and $\alpha(i)$ is the largest positive root of the equation

$$x^i = x^{i-1} + \cdots + x + 1,$$

and s_k^i is of similar type. Therefore, ignoring additive constants,

$$f^i(m, n) \approx \log_{\alpha(i)} m + (i + 1)n \tag{4.3}$$

for $i \geqslant 2$. It is also known that

$$\phi = \frac{1}{2}(1 + \sqrt{5}) = \alpha(2) < \alpha(3) < \cdots \to 2.$$

The formula (4.3) provides additional evidence for the validity of an estimate close to $\log_2 m + (n - 1)\log_2\log_2 m$ for the scheme proposed in 6.3.

6.5   Implementation on a parallel machine

In practice, the number s of rotations that can be performed simultaneously is limited by the number of parallel processors available, and if m is large, then s will be smaller than the theoretical maximum $\lfloor \frac{1}{2}m \rfloor$.

When m is large, the following modification to the scheme described in Section 6.3 is proposed. For the purpose of this brief exposition, we assume that $m \gg s$, as is often likely to be the case.

At the first stage, we annihilate simultaneously the s elements in positions

$$(m, 1),..., (m - s + 1, 1)$$

(by rotating between rows $m,..., m - s + 1$ and $m - s,..., m - 2s + 1$ respectively); at the second stage, we annihilate the elements in positions

$$(m - s, 1),..., (m - 2s + 1, 1);$$

and so on, until this is no longer possible because the number of

Table 5.1   Number of steps for matrices of size
$mn \times n$ using the new scheme, with $s = \frac{1}{2}n$

m	n:	4	8	16
2		12(11)	25(23)	50(47)
3		20(19)	41(39)	82(79)
4		28(27)	57(55)	114(111)
5		36(35)	73(71)	146(143)

rows above is smaller than s. Then we annihilate as many further elements as possible in the first column—this will be a number $s_1 < s$—together with $s - s_1$ elements at the foot of the second column, and continue in this way until the first column is dealt with.

In subsequent stages, s elements in the second column are annihilated simultaneously, until this is no longer possible, and then elements in the second and third columns, and so on.

At each stage, except near the end, s new zeros can be introduced. The scheme is essentially optimal except for edge effects, and the details of implementation would depend on the particular computer architecture.

In Table 5.1 we indicate the number of steps taken for matrices of size $mn \times n$, with $n = 4, 8, 16$ and $m = 2, 3, 4, 5$. The minimum number of steps in this case, with $s = \frac{1}{2}n$, is $[mn^2 - \frac{1}{2}n(n + 1)]/\frac{1}{2}n$ which is $(2m - 1)n - 1$. The figures in parentheses indicate the minimum number possible. Figure 5.1 shows orderings generated for certain values of m and n, with $s = \frac{1}{2}n$, i.e. with an array of $n \times n$ processors.

General notation

It is possible to establish some general notation for expressing Givens sequences: for example, from Fig. 2.1(ii), the set of independent rotations at the first step is

$$\{Q_{16,1}, Q_{15,1}, Q_{14,1}, \ldots, Q_{9,1}\}$$

which can be represented as $Q_{16-9,1}^{(1)}$. The set of independent rotations forms a 'compound rotation,' and the problem can therefore be viewed as essentially a serial one with compound

(a) 20 × 4:

```
 0  0  0  0
10  0  0  0
 9 19  0  0
 9 18 28  0
 8 18 27 36
 8 17 26 35
 7 17 26 34
 7 16 25 34
 6 16 25 33
 6 15 24 33
 5 15 24 32
 5 14 23 32
 4 14 23 31
 4 13 22 31
 3 13 22 30
 3 12 21 30
 2 12 21 29
 2 11 20 29
 1 11 20 28
 1 10 19 27
```

(b) 40 × 8:

```
 0  0  0  0  0  0  0  0
11  0  0  0  0  0  0  0
10 21  0  0  0  0  0  0
10 20 30  0  0  0  0  0
 9 19 29 39  0  0  0  0
 9 19 29 38 48  0  0  0
 9 19 28 38 47 56  0  0
 9 18 28 37 46 55 65  0
 8 18 28 37 46 55 64 73
 8 18 27 37 46 54 63 72
 8 18 27 36 45 54 63 71
 8 17 27 36 45 54 62 71
 7 17 27 36 45 54 62 70
 7 17 26 36 45 53 62 70
 7 17 26 35 44 53 62 70
 7 16 26 35 44 53 61 70
 6 16 26 35 44 53 61 69
 6 16 25 35 44 52 61 69
 6 16 25 34 43 52 61 69
 6 15 25 34 43 52 60 69
 5 15 25 34 43 52 60 68
 5 15 24 34 43 51 60 68
 5 15 24 33 42 51 60 68
 5 14 24 33 42 51 59 68
 4 14 24 33 42 51 59 67
 4 14 23 33 42 50 59 67
 4 14 23 32 41 50 59 67
 4 13 23 32 41 50 58 67
 3 13 23 32 41 50 58 66
 3 13 22 32 41 49 58 66
 3 13 22 31 40 49 58 66
 3 12 22 31 40 49 57 66
 2 12 22 31 40 49 57 65
 2 12 21 31 40 48 57 65
 2 12 21 30 39 48 57 65
 2 11 21 30 39 48 56 64
 1 11 20 30 39 47 56 64
 1 11 20 29 38 47 56 64
 1 10 20 29 38 47 55 63
 1 10 19 28 37 46 55 63
```

Fig. 5.1 Annihilation orderings generated from the new scheme with $s = \frac{1}{2}n$. The matrices are of dimensions (a) 20×4 and (b) 40×8.

transformation. The concept of a Givens sequence can be extended to compound Givens sequences.

Error bounds

For details of implementation of various Givens orderings and timings on the DAP, the reader is referred to Modi and Bowgen

(1984). An added advantage of the new scheme is that the error bounds turn out to be sharper than those of the standard Givens ordering.

It has been shown by Gentleman (1975, Lemma 2) that

if a sequence of Givens transformations can be written as a sequence of s stages, where each stage consists of the simultaneous application of disjoint Givens transformations, then the final computed vector obtained when this sequence of Givens transformations is applied to a given vector u will be the exact result of exact computations on a vector whose difference from u is bounded in norm by $\eta s(1 + \eta)^{s-1} \|u\|$, where η is some fixed constant.

Summarizing the results of Sections 6.2–6.4, we have:

$s = m + n - 2 \ (m > n)$ for the standard scheme,

$s \sim \log_2 m + (n - 1)\log_2\log_2 m \ (m \to \infty)$ for the new scheme,

$s \sim \log_{\alpha(i)} m + (i + 1)n \ (m \to \infty)$ for the Fibonacci schemes.

These results indicate that the schemes of Sections 6.3 and 6.4 are numerically stable.

6.6 Square-root-free Givens transformations

The use of Givens rotations for QR factorization has been encouraged on serial machines by the development of the so-called 'square-root-free' Givens transformation. Originally introduced by Gentleman (1973), this technique reduces considerably the work required and removes the need for the calculation of any square roots.

Let A be an $m \times n$ matrix, with $m \geqslant n$. We construct a nonsingular $m \times m$ matrix M such that

   (i) MA is upper triangular,
   (ii) MM^T is positive diagonal.

Let $MM^T = \text{diag}(\mu_1^2,\ldots, \mu_m^2)$, with $\mu_i > 0 \ (i = 1,\ldots, m)$. Define $(MM^T)^{-\frac{1}{2}} = \text{diag}(\mu_1^{-1},\ldots, \mu_m^{-1})$. Then

$$Q = M^T(MM^T)^{-\frac{1}{2}}$$

is orthogonal; also $R = (MM^T)^{-\frac{1}{2}}MA$ is upper triangular, since

MA is upper triangular and $(MM^\mathsf{T})^{-\frac{1}{2}}$ is diagonal. Thus

$$A = QR$$

is the QR factorization of A.

To construct M, we perform a sequence of recursive multiplications:

$$A_0 = A, \qquad A_k = K_k A_{k-1} \quad (k = 1, \ldots, r) \qquad r = \tfrac{1}{2}n(2m - n - 1),$$

where the $m \times m$ matrices K_k are defined below. Let

$$M_0 = I_m \qquad M_k^\mathsf{T} = K_1^\mathsf{T} \cdots K_k^\mathsf{T} \quad (k = 0, \ldots, r),$$

so that

$$A_k = M_k A.$$

Here r is the number of subdiagonal elements of A. Then, for $k = 0, \ldots, r - 1$ the matrices are constructed to have the following properties.

   (i)′ The multiplication by K_k preserves, in A_k, the null subdiagonal elements of A_{k-1}, while it creates a zero, in A_k, in the same position as a selected nonzero subdiagonal element of A_{k-1}.
   (ii)′ $M_k M_k^\mathsf{T}$ is positive diagonal.
   (iii)′ Each transformation $X \leftarrow K_k X$ affects just two rows of X.

We annihilate downwards from the diagonal, one element at a time, in the order

$$(i, j) < (i', j') \quad \Leftrightarrow \quad i < i' \vee (i = i' \wedge j < j').$$

Suppose that, on iteration k (with $k \geqslant 1$), we have to annihilate the (p, q) element, with $p > q$. Let x_1 and x_2 be respectively the (q, q) and (p, q) elements of A_{k-1}. If $(x_1, x_2) = (0, 0)$, set $K_k = I$. If $(x_1, x_2) \neq (0, 0)$, define K_k to be the matrix whose $\{p, q\} \times \{p, q\}$ submatrix Λ is defined as follows, with $(K_k)_{i,j} = \delta_{ij}$ for $\{i, j\} \nsubseteq \{p, q\}$.

$$\Lambda = \begin{cases} \Lambda_1 = \begin{bmatrix} d_2 x_1 / d_1 x_2 & 1 \\ 1 & -x_1 / x_2 \end{bmatrix} & \text{if } x_2 \neq 0 \text{ and } d_2 x_1^2 / d_1 x_2^2 < 1, \\[2ex] \Lambda_2 = \begin{bmatrix} 1 & d_1 x_2 / d_2 x_1 \\ -x_2 / x_1 & 1 \end{bmatrix} & \text{if } x_1 \neq 0 \text{ and } d_1 x_2^2 / d_2 x_1^2 \leqslant 1, \end{cases}$$

where $D = \text{diag}(d_1, d_2)$ is the $\{p, q\} \times \{p, q\}$ submatrix of $M_{k-1}M_{k-1}^\mathsf{T}$. It is straightforward to show that K_k satisfies (i)′, and (ii)′ can be shown by a simple inductive argument. Thus, after r steps (at most), we obtain (i) and (ii) from (i)′ and (ii)′, where

$$M = M_r.$$

We have

$$\Lambda_1 D\Lambda_1^\mathsf{T} = (1 + \gamma_1)\,\text{diag}(d_2, d_1) \quad (x_2 \neq 0),$$
$$\Lambda_2 D\Lambda_2^\mathsf{T} = (1 + \gamma_2)\,\text{diag}(d_2, d_1) \quad (x_1 \neq 0),$$

where $\gamma_1 = d_2 x_1^2/d_1 x_2^2$ and $\gamma_2 = d_1 x_2^2/d_2 x_1^2$. By the choice of Λ, the 'amplification' factor $1 + \gamma_1$ or $1 + \gamma_2$ always lies in the range $[1, 2]$.

Matrices of the form

$$\Lambda_1 = \begin{bmatrix} \beta_1 & 1 \\ 1 & \alpha_1 \end{bmatrix}, \qquad \Lambda_2 = \begin{bmatrix} 1 & \alpha_2 \\ \beta_2 & 1 \end{bmatrix}, \tag{6.1}$$

with $-1 \leq \alpha_i\beta_i \leq 0$, are referred to as *fast*, or as square-root-free, Givens matrices, since the calculation of coefficients requires about half the multiplications of a conventional Givens (Table 6.1), and does not involve a square root.

The arithmetic operation counts for both standard and square-root-free transformations are shown in Table 6.1. The saving in using the square-root-free transformation clearly comes from the matrix-product phase which effectively swamps the calculation of the rotation angles.

On a parallel machine such as the DAP, the calculation of the rotation angle with the standard method costs about the same as the product phase (for $n \leq 64$) which entails two matrix products.

Table 6.1 Operations for calculating a single Givens rotation

	Rotation angles				Matrix product	
	$\uparrow^2$	$\sqrt{\,}$	$\div$	$+$	$*$	$+$
Standard	2	1	2	1	$4n$	$2n$
Square-root-free	$\uparrow^2$	$*$	$\div$	$+$	$*$	$+$
	2	4	3	1	$2n$	$2n$

Therefore we would not expect such great savings using square-root-free transformations as can be obtained on a serial computer. Indeed, on the DAP, the calculation of the rotation angles for the square-root-free version takes longer than the standard case. The organizational overheads for the square-root-free DAP code are considerable and almost neutralize the saving of having one less matrix multiplication. The rotation-angle and product phases take about 1 ms and 0.7 ms respectively.

6.7 QR factorization via Householder reflections

The theory behind Householder reduction is well known. We describe this briefly and fix our notation. Let us assume that A is a real $m \times n$ matrix, with $m \geq n$, which is to be expressed as QR, where Q is an $m \times m$ matrix and $R = \begin{bmatrix} \tilde{U} \\ 0 \end{bmatrix}$ with $\tilde{U}$ an $n \times n$ upper triangular matrix.

A sequence of elementary reflectors $P^{(0)}, \ldots, P^{(q)}$ is applied on the left of A such that the product $R = P^{(q)} \cdots P^{(1)} P^{(0)} A$ is upper trapezoidal, where

$$q = \begin{cases} n - 2 & (m = n), \\ n - 1 & (m > n). \end{cases}$$

The orthogonal matrices $P^{(r)}$ $(r = 0, \ldots, q)$ are of the form $P^{(r)} = I - 2w^{(r)}w^{(r)\mathsf{T}}$, where $\|w^{(r)}\|_2 = 1$, and $w^{(r)} = [0, 0, \ldots, w_{r+1}^{(r)}, \ldots, w_m^{(r)}]^\mathsf{T}$. We take

$$w^{(r)} = h^{(r)} / (2\sigma^2 + 2|u_{r+1}|\sigma)^{\frac{1}{2}},$$

where $h^{(r)} = [0, 0, \ldots, (\operatorname{sgn} u_{r+1})(\sigma + |u_{r+1}|), u_{r+2}, \ldots, u_m]^\mathsf{T}$, with $u_j = a_{j,r+1}^{(r+1)}$ and

$$\sigma = \left(\sum_{j=r+1}^{m} |u_j|^2 \right)^{\frac{1}{2}}.$$

Application of $P^{(r)}$ introduces zeros below the main diagonal of A in column $r + 1$. The matrix Q is recovered by forming the product $P^{(0)} \cdots P^{(q)}$. The details of the algorithm can be found in Stewart (1973).

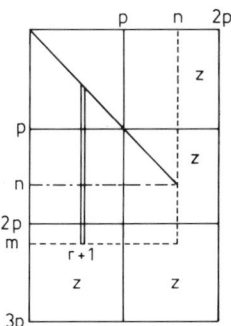

Fig. 7.1 QR factorization of an $m \times n$ matrix mapped onto a $p \times p$ array of processors

Implementation

Suppose we have a $p \times p$ array of processors and a matrix A of size $m \times n$, with $m \geq n \geq p$. The implementation is quite simple once we map the problem on the $p \times p$ array of processors, as illustrated in Fig. 7.1.

The matrix bordered by a sufficient number of zeros is sliced in the obvious way into a matrix of $M \times N$ blocks of size $p \times p$. Areas marked z in Fig. 7.1 are set to zero by the routine. The algorithm then proceeds as follows.

1. Set $r = 0$.
2. Set $\boldsymbol{u} = $ column $r + 1$ of A.
3. Calculate $\sigma = (\sum_{r+1}^{m} |u_j|^2)^{\frac{1}{2}}$ from $\boldsymbol{u}$.
4. Calculate $\rho = \sigma(\sigma + |u_{r+1}|)$ and set

   $$u_{r+1} \leftarrow (\operatorname{sgn} u_{r+1})(\sigma + |u_{r+1}|).$$

5. If $\rho = 0$, go to 9.
6. Evaluate $\boldsymbol{v} = A^{\mathsf{T}}\boldsymbol{u}/\rho$ for the first column of $M \times N$ blocks.
7. Update the first column of $M \times N$ blocks by $A \leftarrow A - \boldsymbol{u}\boldsymbol{v}^{\mathsf{T}}$.
8. Repeat 6–7 for block columns 2 to N if necessary.
9. Set $r \leftarrow r + 1$.
10. If $r < n$, go to 2.

The algorithm is serial over the columns of A but all the vector and matrix operations in Steps 6 and 7 are performed in parallel.

Table 7.1   DAP timings for QR factorization (seconds)

M	N	Householder time	Givens time	Ratio Givens/Householder
8	8	16.6	26.4	1.59
4	4	2.75	4.4	1.60
2	2	0.556	0.88	1.58
1	1	0.145	0.19	1.31
2	1	0.235	0.44	1.87
4	1	0.411	0.88	2.14
8	1	0.762	1.76	2.31

A is transformed into $R,$ and Q can be recovered by using

$$Q = P^{(0)}P^{(1)}\cdots P^{(q)}.$$

This turns out to be faster, on the DAP, than the Givens method. Table 7.1 gives the DAP timings on a 64×64 array of processors, and compares them with those obtained using the standard Givens sequence. The improvement is more significant as M becomes larger than N. For further details the reader is referred to Bowgen and Modi (1985).

Additional notes

Cosnard and Robert (1986) have recently carried out some theoretical studies on optimal Givens sequences. They prove that the new scheme developed by Modi and Clarke (Modi 1982, Modi and Clarke 1984) is always optimal, not only in the case of Givens sequences, but, more generally, with respect to all possible parallel orderings. The formula for the number of parallel steps (see Section 6.3) can therefore be viewed as a result of complexity. Cosnard and Robert also show that, when m and n are proportional, the Fibonacci scheme of order 1 is to be preferred to Sameh and Kuck's (1978) scheme, unless the matrix is square.

The QR factorization has been investigated by various authors: systolic arrays by Bojanczyk *et al.* (1984) and Gentleman and Kung (1981), and the square-root-free Givens method by Barlow and Ipsen (1984); data flow architectures by O'Leary and Stewart

(1984); implementation on the DAP by Modi and Bowgen (1984).

Details of the implementation of Sameh's scheme for a mesh-connected array of processors are examined by Gannon (1980). Lord *et al.* (1983) also consider a Givens sequence, with a zigzag pattern of annihilation, and assess its effectiveness on the Deneclor HEP, an MIMD system. Sameh (1985) discusses the use of Givens transformations on a ring of interconnected processors.

7 Singular-value decomposition and related problems

7.1 Introduction

This chapter examines parallel algorithms for accomplishing the singular-value decomposition (SVD) and the generalized singular-value decomposition (GSVD), both techniques being applicable to a wide range of practical problems such as signal processing and statistical analysis.

The SVD decomposes a real $m \times n$ matrix A $(m \geqslant n)$ into $U \begin{bmatrix} \Sigma \\ 0 \end{bmatrix} V^\mathsf{T}$, where U and V are orthogonal matrices of order $m \times m$ and $n \times n$ respectively and Σ is an $n \times n$ diagonal matrix. Early references date back to Kogbetliantz (1954, 1955) and Hestenes (1958), and the methods proposed resemble closely the Jacobi method for the symmetric eigenvalue problem where the decomposition is based on the application of plane rotations. However, these were superseded by the method of Golub and Kahan (1965), who suggested applying Householder transformations to reduce A to bidiagonal form and then using the QR algorithm.

The approach considered here is to apply the parallel Jacobi method discussed in Chapter 5. The method is similar to that of Hestenes, in that we multiply A from the right by orthogonal transformations to diagonalize $A^\mathsf{T}A$. For $m \gg n$, we first perform a QR factorization $A = Q \begin{bmatrix} \bar{U} \\ 0 \end{bmatrix}$, where the $m \times m$ matrix Q is a product of Givens plane rotations, or Householder reflections, and $\bar{U}$ is an $n \times n$ upper triangular matrix. We then proceed to compute the SVD of the much smaller matrix $\bar{U}$ as before.

The GSVD is essentially an extension of the SVD for any two matrices with the same number of columns. In order to facilitate the computation of the generalized eigenvalue problem, which often arises in linear system theory, we also discuss contragredient transformations and SVD for a matrix product. Both SVD and GSVD constitute numerically reliable mathematical tools which are invaluable in solving numerous problems. The SVD application to the linear least-squares problem is discussed in Section 7.4, where we also consider the use of the generalized inverse A^+ of Moore and Penrose. The GSVD provides an efficient solution for the general Gauss–Markov linear model and the associated weighted least-squares problem (Sections 7.5 to 7.7). We briefly describe the Kalman filtering problem (Section 7.9) and its solution via the least-squares approach, and finally (Section 7.10), we indicate how the SVD can be used to solve the standard eigenvalue problem.

7.2 Singular-value decomposition

The singular-value decomposition of a real $m \times n$ matrix A $(m \geq n)$ is formally defined as

$$A = U \begin{bmatrix} \Sigma \\ 0 \end{bmatrix} V^\mathsf{T}, \tag{2.1}$$

where the $m \times m$ matrix U and the $n \times n$ matrix V are orthogonal and

$$\Sigma = \mathrm{diag}(\sigma_1, \ldots, \sigma_n).$$

The columns of U and V respectively consist of m and n orthonormalized eigenvectors associated with m and n eigenvalues of AA^T and $A^\mathsf{T}A$. The non-negative entries in Σ are the singular values of A. Equation (2.1) gives

$$U^\mathsf{T}AV = \begin{bmatrix} \mathrm{diag}(\sigma_1, \ldots, \sigma_n) \\ \mathrm{O}_{(m-n) \times n} \end{bmatrix},$$

$$V^\mathsf{T}(A^\mathsf{T}A)V = \mathrm{diag}(\sigma_1^2, \ldots, \sigma_n^2), \tag{2.2}$$

$$U^\mathsf{T}(AA^\mathsf{T})U = \mathrm{diag}(\sigma_1^2, \ldots, \sigma_n^2) \oplus \mathrm{O}_{(m-n) \times (m-n)}, \tag{2.3}$$

where $P \oplus Q = \begin{bmatrix} P & \mathrm{O} \\ \mathrm{O} & Q \end{bmatrix}$ is the direct sum of P and Q.

Further, if we partition U as

$$U = [U_1, U_2],$$

where U_1 is $m \times n$, and U_2 is $m \times (m - n)$, and let

$$W = \frac{1}{\sqrt{2}} \begin{bmatrix} V & V & 0 \\ U_1 & -U_1 & kU_2 \end{bmatrix},$$

where k is a scalar, then

$$W^\mathsf{T} \begin{bmatrix} 0 & A^\mathsf{T} \\ A & 0 \end{bmatrix} W = \mathrm{diag}(\sigma_1,\ldots,\sigma_n, -\sigma_1,\ldots, -\sigma_n) \oplus O_{(m-n)\times(m-n)}.$$

$$(2.4)$$

Equations (2.2), (2.3), and (2.4) show the relationship with the eigenvalue problem discussed in Chapter 5 (for further details see Golub and Van Loan 1983).

The definition of SVD for the case of complex matrices generalizes in the obvious way.

7.3 Singular-value decomposition via Jacobi rotations

Applying the Jacobi method outlined in Chapter 5, we perform plane rotations $P^{(k)}$ on $B = A^\mathsf{T}A$, where A is $m \times n$ and B is $n \times n$, so that $\tan 2\theta = 2b_{ij}/(b_{ii} - b_{jj})$, with θ in the range $-\frac{1}{4}\pi < \theta \leq \frac{1}{4}\pi$. Formally we have

$$P^{(s)} \cdots P^{(1)} B P^{(1)\mathsf{T}} \cdots P^{(s)\mathsf{T}} \equiv P^{(s)} \cdots P^{(1)} A^\mathsf{T} A P^{(1)\mathsf{T}} \cdots P^{(s)\mathsf{T}},$$

where s is the final iteration step. If we accumulate the product $P^{(1)} \cdots P^{(s)} = V$, then we have

$$U \begin{bmatrix} \Sigma \\ 0 \end{bmatrix} = A P^{(1)} \cdots P^{(s)},$$

hence the SVD of A. However, the condition number $\mathrm{cond}(P^{(s)} \cdots P^{(1)} A^\mathsf{T} A P^{(1)\mathsf{T}} \cdots P^{(s)\mathsf{T}})$ is the square of $\mathrm{cond}\,A$, and this can lead to serious loss of accuracy. An alternative approach is as follows.

Given A, form the sequence of plane rotations $P^{(1)},\ldots, P^{(s)}$ determined by

$$\tan 2\theta = 2a_i^\mathsf{T} a_j/(a_i^\mathsf{T} a_i - a_j^\mathsf{T} a_j); \qquad (3.1)$$

then

$$U \begin{bmatrix} \Sigma \\ 0 \end{bmatrix} = A P^{(1)} \cdots P^{(s)}. \qquad (3.2)$$

We can derive the result given by (3.1) in the following way. Consider the action of a plane rotation on the right-hand side of A which affects two columns, for example a_i and a_j.

Then

$$[a_i, a_j]\begin{bmatrix} c & -s \\ s & c \end{bmatrix} = [a_i', a_j'],$$

with, as always, $c = \cos \theta$ and $s = \sin \theta$. Thus

$$a_i' = a_i c + a_j s, \qquad a_j' = -a_i s + a_j c.$$

For orthogonality, we require $a_i'^{\mathrm{T}} a_j' = a_j'^{\mathrm{T}} a_i' = 0$, which gives

$$(-a_i^{\mathrm{T}} a_i + a_j^{\mathrm{T}} a_j)cs + a_i^{\mathrm{T}} a_j(c^2 - s^2) = 0,$$

and hence we obtain (3.1).

Alternatively, we may wish to choose θ so that the euclidean norm of a_i', i.e. $\|a_i'\|_2$, is maximized (Nash 1975). This is equivalent to maximizing

$$y = \|a_i'\|_2^2 = a_i'^{\mathrm{T}} a_i'.$$

Differentiating with respect to θ gives $dy/d\theta = 2a_i'^{\mathrm{T}} a_j'$, so that a turning value occurs when $a_i'^{\mathrm{T}} a_j' = 0$, giving (3.1).

Further, for the singular values to be ordered, we require the second-order condition $d^2y/d\theta^2 < 0$, which gives

$$\operatorname{sgn} t = \operatorname{sgn} a_i^{\mathrm{T}} a_j \quad \text{for } a_i^{\mathrm{T}} a_j \neq 0,$$

with $t = \tan \theta$, and

$$c = \begin{cases} 1 & \text{when } a_i^{\mathrm{T}} a_i \geq a_j^{\mathrm{T}} a_j \\ 0 & \text{when } a_i^{\mathrm{T}} a_j < a_j^{\mathrm{T}} a_j \end{cases} \quad \text{for } a_i^{\mathrm{T}} a_j = 0.$$

The second-order condition ensures that $a_i'^{\mathrm{T}} a_i' - a_i^{\mathrm{T}} a_i \geq 0$, so that calculation always proceeds towards an ordering of column norms.

Assuming (without loss of generality) that n is even, and applying simultaneous rotations—as for the Jacobi method in Chapter 5—we obtain $\frac{1}{2}n$ pairs of orthogonalized columns at each step. The process is iterative, because the orthogonality established at previous steps is subsequently destroyed. However, the amount of nonorthogonality is sufficiently reduced at each step.

To see this, define a measure of nonorthogonality as

$$\sigma^{(k)} = \sum_{i=1}^{n-1} \sum_{j=i+1}^{n} (a_i^{(k)\mathsf{T}} a_j^{(k)})^2;$$

then, since

$$[(cx + sy)^{\mathsf{T}} z]^2 + [(-sx + cy)^{\mathsf{T}} z]^2 \equiv (x^{\mathsf{T}} z)^2 + (y^{\mathsf{T}} z)^2,$$

we have $\sigma^{(k)} = \sigma^{(k-1)} - x$, where $x > 0$ is the sum of the squares of the inner products of the $\tfrac{1}{2}n$ pairs of columns orthogonalized at the kth step.

The parallel SVD algorithm is thus defined as follows.

Let $Z = \{(p, q) : 1 \leqslant p < q \leqslant n\}$ (here p and q indicate columns of A); partition Z so that

$$|Z_k| = \tfrac{1}{2}n \quad (k = 1,\ldots, n-1).$$

Set $A^{(1)} = A$, $V^{(1)} = I_n$, and, for $k = 1,\ldots, n-1$, compute

$$A^{(k+1)} = A^{(k)} P^{(k)}, \qquad V^{(k+1)} = V^{(k)} P^{(k)}.$$

After completing $n - 1$ iterations (one sweep), the process is repeated with the initial A replaced by the latest $A^{(k+1)}$ computed in the previous sweep, until the columns of A are orthogonalized to the required degree of accuracy. Then

$$U \begin{bmatrix} \Sigma \\ 0 \end{bmatrix} = A^{(s)}, \qquad V = V^{(s)},$$

where s is the final iteration step.

Here, the matrix $P^{(k)}$ is defined by

$$\left.\begin{array}{l} \alpha^{(k)}(p, q) = \tfrac{1}{2}\tan^{-1}[2a_i^{\mathsf{T}} a_j/(a_i^{\mathsf{T}} a_i - a_j^{\mathsf{T}} a_j)], \\[4pt] P^{(k)}(p, q) = -P^{(k)}(q, p) = \sin \alpha^{(k)}(p, q), \\[4pt] P^{(k)}(p, p) = P^{(k)}(q, q) = \cos \alpha^{(k)}(p, q), \end{array}\right\} \quad \text{for } (p, q) \in Z_k,$$

with $P^{(k)}(i, j) = \delta_{ij}$ for each (i, j) such that:

$$\{i, j\} \nsubseteq \{p, q\} \quad \text{for all } (p, q) \in Z_k.$$

(The superscripts k on the columns of matrix $A^{(k)}$ are dropped for clarity.)

The method is essentially the same as the parallel Jacobi method. At each step, instead of diagonalizing a number of 2×2

matrices (as for the symmetric eigenvalue problem), we perform a number of independent SVDs simultaneously. There are a number of ways in which the set Z_k can be chosen, and the choice is strongly dependent upon the data movement involved (Chapter 5, Sections 5.6 and 5.7).

In comparison, Kogbetliantz's method consists in generating a sequence of matrices $A^{(k)}$ as follows. Set

$$A^{(1)} = A, \qquad U^{(1)} = I_m, \qquad V^{(1)} = I_n,$$

and, for $k = 1, 2, \ldots$, compute

$$A^{(k+1)} = U_k A^{(k)} V_k^T, \qquad U^{(k+1)} = U_k U^{(k)}, \qquad V^{(k+1)} = V_k V^{(k)},$$

so that A converges to a diagonal matrix. The unitary updating transformations U_k and V_k are chosen to be complex elementary rotations through angles ϕ_k and ψ_k which annihilate the elements of $A^{(k)}$ in the symmetric positions (i_k, j_k) and (j_k, i_k). The method is similar to the Jacobi method, and indeed is identical for the case of symmetric matrix A. The iterations are performed so as to reduce eventually the square root of the sum of squares of off-diagonal elements

$$\text{off } A = \left(\sum_{i \neq j} |a_{ij}|^2 \right)^{\frac{1}{2}}$$

to zero to machine precision. In practice, off A is observed to decrease *quadratically*, and this is proved by Paige and Van Dooren (1986) for the case when A has no repeated singular values, and by Charlier and Van Dooren (1986) in the general case. For a general $m \times n$ matrix A with $m > n$, a saving in time and storage may be obtained by performing a preliminary QR factorization of A, as explained in Step 1 below. Much of the discussion on the parallel Jacobi method (Chapter 5), such as the simultaneous application of rotations, the use of mobile schemes, and the application of approximate rotations, readily applies to Kogbetliantz's method. Even the proof of quadratic convergence is strongly influenced by the corresponding proof for the cyclic Jacobi method.

In Table 3.1, comparison is made between the DAP timings for the parallel SVD and a similar algorithm implemented on the Illiac IV (Luk 1980).

It is seen that, for higher-order matrices, the DAP should

Table 3.1 Comparison of results for a matrix of order 64 on the DAP
and the Illiac IV

	Sweeps	Time (sec)	Order of the algorithm ($m = n$)	Algorithm
DAP	6	1.5	$O(n)$	Jacobi type
Illiac IV	9	4.56	$O(n^2)$	Jacobi type
ICL 2980		7.3	$O(n^3)$	Householder + QR algorithm

perform better than the Illiac IV. This is because the Illiac IV
only executes vector operations of order n in parallel, whereas
the DAP executes matrix operations of order n^2 in parallel,
provided that sufficient processors are available.

Numerical test results

Table 3.2 gives the singular values obtained for the matrix

$$W = \begin{bmatrix} 1 & & & \\ & 1 & \cdot & -1 \\ & & \cdot & \cdot \\ 0 & & & \cdot \\ & & & 1 \end{bmatrix}$$

of order 64, which is ill conditioned, since it has a very small
singular value. The matrix becomes singular if we add -2^{-n+2} to
its $(n, 1)$ position. The results are consistent to machine precision
with the NAG routine (Mk8 version) run on the ICL 2980. Error
analysis and convergence for the parallel Jacobi method (Chapter
5, Section 5.9) indicate the stability of the method.

In the above algorithm, iterations were performed on the
entire plane of A. We modify the algorithm so that it first
reduces A to a smaller size, and is particularly suited when
$m \gg n$.

For a rectangular $m \times n$ matrix A, with $m \gg n$, the algorithm is
defined in two steps.

Step 1 Perform a QR factorization $A = Q_1 R$, where Q_1 is an
$m \times m$ orthogonal matrix, so that $Q_1^T Q_1 = I_m$, and $\bar{U}$ is an $n \times n$

Table 3.2 Singular values of the *W*-matrix of order 64

0.15379552332590048E-06	1.500455746764526	1.500202161059790	1.886253227858382
1.50081242624241383	2.154030578873773	1.501273878583472	1.646470871192256
1.502520380764783	2.387125237611773	1.510693308372752	1.620434363710420
1.514316545827085	3.024458532614274	1.518643904376440	1.599293369669204
1.507692205651664	3.898293385968094	1.537025871250081	1.59019997176024
1.500500481459815	4.653879383994490	1.512422481914019	1.744829570078014
1.509119401652127	8.089530911279976	1.504219283098848	1.996955342457639
1.516386309361169	1.678947616152306	1.506404129556839	1.661785239008689
1.550327972152809	2.068267890254135	1.52668350440256	2.258391583391316
1.533288846357759	1.720070311691490	1.505248459625462	2.54848620028221
1.561307597917457	1.698255901755135	1.567569776693719	1.773072225321787
1.50311422743904	3.388177368081026	1.555577833326836	2.75467987896880
1.523778706735915	1.937044877155591	1.545512081094040	1.805465041317393
1.581941322949759	5.867138017606183	1.541089863262936	1.632756500104921
1.521102944002073	13.34241207501316	1.539851161598285	1.842841301252664
1.501842291003758	1.609329413432777	1.574424588237554	39.81530867204764

upper triangular matrix:

$$A = Q_1 \begin{bmatrix} \bar{U} \\ 0 \end{bmatrix}$$

The parallel QR factorization algorithm developed in Chapter 6 may be used here.

Step 2 Find the SVD of $\bar{U}$:

$\bar{U} = Q_2 \Sigma P^{\mathsf{T}}$, where $\Sigma = \mathrm{diag}(\sigma_1, \dots, \sigma_n)$ with $\sigma_1 \geqslant \cdots \geqslant \sigma_n \geqslant 0$.

The SVD of A is therefore

$$A = Q_1 \begin{bmatrix} Q_2 & 0 \\ 0 & I \end{bmatrix} \begin{bmatrix} \Sigma \\ 0 \end{bmatrix} P^{\mathsf{T}}.$$

For Step 2, we need to find the SVD of an $n \times n$ matrix, and for this we can use the earlier algorithm. In many practical situations, such as those involving least squares, we should check the numerical rank of $\bar{U}$ and proceed to Step 2 only where necessary.

Numerical properties and timings

Figure 3.1 gives the timings for the parallel method for $64l \times 64$ matrices $(l = 2, \dots, 10)$ and compares them with the NAG routine.

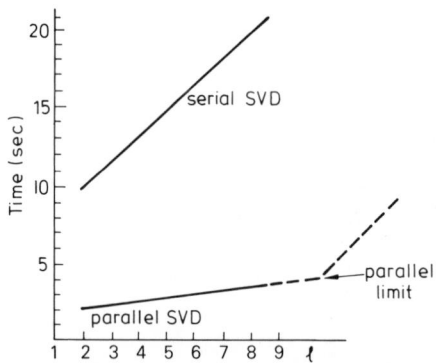

Fig. 3.1 Timing comparison for the parallel SVD algorithm with the NAG routine for $64l \times 64$ matrices. The NAG routine used was the Mk 8 version, based on Householder transformations and the QR algorithm, and run on an ICL 2980.

The algorithm developed on the DAP can take any value of m and n, not necessarily a multiple of 64. However, if m is not a multiple of 64 or if $n \le 64$, there will be some redundancy in fully utilizing all the processing elements. If $n > 64$, then it becomes necessary to partition the matrix into smaller, manageable units. This can be done by slicing or crinkling (see Chapter 2, Section 2.9).

For matrices of order $64l \times 64$, we can estimate the timings as follows. Each 64×64 block takes approximately 0.3 sec, so the time for Step 1 is $0.3l$ sec. Step 2 consists in finding the SVD of a 64×64 matrix, which takes 1.5 sec for six sweeps, and hence the total time, in single-precision arithmetic, is $(0.3l + 1.5)$ sec.

In the above experiments, we used the QR factorization based on Givens rotations. We could equally have used the Householder reflections discussed in Chapter 6 (Section 6.7).

Approximate formulae

It was observed earlier (in connection with the Jacobi method) that, on some parallel systems, the computational cost is dominated by the arithmetic operations in the main loop of the algorithm, which computes the coefficients $s = \sin \alpha$ and $c = \cos \alpha$ in the plane rotations and performs the matrix multiplication. The relative cost of computing (c, s) is much greater on a parallel system than on a serial machine. The approximate formulae devised in Chapter 5 (Section 5.10) might therefore be applied in order to increase computation speed. A recent paper by Charlier *et al.* (1986) considers the use of these formulae for SVD algorithms and demonstrates an improvement in speed not only on parallel systems but also on serial machines. The authors extend the use of approximate formulae to the hermitian case, as well as carrying out an error analysis.

7.4 The linear least-squares problem

We consider the problem of determining an $n \times 1$ vector x which minimizes the function

$$\|r\|_2 = \|b - Ax\|_2, \tag{4.1}$$

where A is an $m \times n$ matrix and b and r are $m \times 1$. The phrase *linear least-squares problem* is used, since the sum of squares of the residual vector $r = b - Ax$ is to be minimized. This problem arises in a number of areas and is clearly related to the approximation of data; in statistics, for example, it arises in regression and multivariate analysis.

If we let $r = b - Ax$, then $r^T r = (b - Ax)^T(b - Ax)$; differentiating this with respect to x and equating the derivative to zero gives the system of *normal equations*:

$$A^T A x = A^T b,$$

where $A^T A$ is an $n \times n$ matrix. We can solve for x using Choleski factorization or Gaussian elimination. However, the condition number of $A^T A$ is the square of that of A, and the equations may therefore be ill conditioned. The following method, which uses QR factorization and SVD, provides a stable and practical way of solving (4.1) without having to compute $A^T A$. First perform a QR factorization of A:

$$A = Q \begin{bmatrix} \bar{U} \\ 0 \end{bmatrix}, \qquad Q^T Q = I_m,$$

where Q is $m \times m$, and $\bar{U}$ is $n \times n$ upper triangular. Substituting into (4.1) gives

$$\|r\|_2 = \left\| Q \left(Q^T b - \begin{bmatrix} \bar{U} \\ 0 \end{bmatrix} x \right) \right\|_2,$$

i.e.

$$\|r\|_2 = \left\| Q^T b - \begin{bmatrix} \bar{U} \\ 0 \end{bmatrix} x \right\|_2.$$

(The 2-norm is unchanged by an orthogonal transformation.) We can write this as

$$\|r\|_2 = \left\| \begin{bmatrix} v \\ w \end{bmatrix} - \begin{bmatrix} \bar{U}x \\ 0 \end{bmatrix} \right\|_2, \qquad Q^T b = \begin{bmatrix} v \\ w \end{bmatrix},$$

where v and w are vectors of n and $m - n$ elements respectively.

Hence

$$\|r\|_2 = \left\| \begin{matrix} \bar{v} - \bar{U}x \\ w \end{matrix} \right\|_2. \tag{4.2}$$

If $\bar{U}$ is nonsingular, then we can solve for x using $v = \bar{U}x$ and

$\|r\|_2 = \|w\|_2$. In order to check the nonsingularity of $\bar{U}$, we can use the condition-number estimate as in the Linpack-Users Guide, and in Cline *et al.* (1979). If $\bar{U}$ is nearly singular, then we find the SVD of $\bar{U}$:

$$\bar{U} = Q_2 \Sigma P^\mathsf{T}, \qquad Q_2^\mathsf{T} Q_2 = I_n, \qquad P^\mathsf{T} P = I_n,$$

where $\bar{U}$, Q_2, Σ, and P^T are all $n \times n$ matrices, and

$$\Sigma = \text{diag}(\sigma_1, \ldots, \sigma_r, 0, \ldots, 0) \quad \text{with } r < n.$$

Substituting into (4.2) leads to

$$\|r\|_2 = \left\| \begin{matrix} v - Q_2 \Sigma P^\mathsf{T} x \\ w \end{matrix} \right\|_2 = \left\| \begin{matrix} Q_2(Q_2^\mathsf{T} v - \Sigma P^\mathsf{T} x) \\ w \end{matrix} \right\|_2$$

$$= \left\| \begin{bmatrix} Q_2 & 0 \\ 0 & I \end{bmatrix} \begin{bmatrix} Q_2^\mathsf{T} v - \Sigma P^\mathsf{T} x \\ w \end{bmatrix} \right\|_2.$$

Hence

$$\|r\|_2 = \left\| \begin{matrix} Q_2^\mathsf{T} v - \Sigma P^\mathsf{T} x \\ w \end{matrix} \right\|_2 = \left\| \begin{bmatrix} z \\ w \end{bmatrix} - \begin{bmatrix} \Sigma y \\ 0 \end{bmatrix} \right\|_2,$$

where $z = Q_2^\mathsf{T} v$, $y = P^\mathsf{T} x$. If we now let

$$z = \begin{bmatrix} \bar{z} \\ \hat{z} \end{bmatrix},$$

where $\bar{z}$ is a vector of $r < n$ elements, and $\hat{z}$ has $n - r$ elements, then

$$\|r\|_2 = \left\| \begin{bmatrix} \bar{z} \\ \hat{z} \\ w \end{bmatrix} - \begin{bmatrix} D & 0 \\ 0 & 0 \end{bmatrix} \begin{bmatrix} \bar{y} \\ \hat{y} \end{bmatrix} \right\|_2,$$

where

$$y = \begin{bmatrix} \bar{y} \\ \hat{y} \end{bmatrix}, \qquad D = \text{diag}(\sigma_1, \ldots, \sigma_r),$$

that is,

$$\|r\|_2 = \left\| \begin{matrix} \bar{z} - D\bar{y} \\ \hat{z} \\ w \end{matrix} \right\|_2.$$

Hence we can solve $\bar{z} = D\bar{y}$ for $\bar{y}$, and, with $\hat{y}$ arbitrary, then form

$$x = Py = P\begin{bmatrix} \bar{y} \\ \hat{y} \end{bmatrix},$$

with $\|r\|_2 = \left\|\begin{matrix} \hat{z} \\ w \end{matrix}\right\|_2$. The solution for which $\|x\|_2 = \|y\|_2$ is a minimum has $\hat{y} = 0$.

Alternatively, we can express the least-squares solution via the pseudoinverse A^+ of A. Originally defined by Moore and Penrose in the 1950s, this later came to be known as the *generalized inverse* of A (Penrose 1955). For the case when A is square and nonsingular, we have $A^+ = A^{-1}$. For an overdetermined system of equations $Ax = b$, with rank $A = n$ (full column rank), and $A^T A$ invertible, the unique solution is $x = (A^T A)^{-1} A^T b$ (and if A^{-1} exists, then $x = A^{-1} A^{-T} A^T b = A^{-1} b$). The pseudoinverse in this case is $A^+ = (A^T A)^{-1} A^T$. However, when A is of rank $r < n$, and $A^T A$ is not invertible, then x is not uniquely determined. The optimal solution is chosen to give minimum $\|x\|_2$ such that $\|b - Ax\|_2$ is minimized, i.e. has the minimum length. The SVD provides an explicit expression for the pseudoinverse as follows.

Equation (4.1) gives $\|r\|_2 = \|b - Ax\|_2$; thus, if the SVD of the $m \times n$ matrix A is $U\begin{bmatrix} \Sigma \\ 0 \end{bmatrix}V^T$, where U and V are orthogonal $m \times m$ and $n \times n$ matrices respectively, and where the diagonal elements of $\Sigma = \text{diag}(\sigma_1, \ldots, \sigma_n)$ are the singular values of A, then we require

$$\left\| b - U\begin{bmatrix} \Sigma \\ 0 \end{bmatrix}V^T x \right\|_2 = 0.$$

Hence $U\begin{bmatrix} \Sigma \\ 0 \end{bmatrix}V^T x = b$, i.e. $x = V\Omega U^T b$, where Ω is the $n \times m$ matrix

$$\Omega = \begin{bmatrix} \text{diag}(\sigma_1^{-1}, \ldots, \sigma_r^{-1}) & O_{r \times (m-r)} \\ O_{(n-r) \times r} & O_{(n-r) \times (m-r)} \end{bmatrix}$$

in which $\sigma_1^{-1}, \ldots, \sigma_r^{-1}$ are the reciprocals of the nonzero singular values in some order; the pseudoinverse of A is $A^+ = V\Omega U^T$, which is independent of this order, so that $x = A^+ b$.

7.5 The general Gauss–Markov linear model

An important use of the least-squares approach is in the numerical solution, for x, of the *general Gauss–Markov linear model* given by

$$b = Ax + r, \tag{5.1}$$

where $r \leftarrow N(0, \sigma^2 W)$, A is an $m \times p$ matrix of observations, b is an m-element vector of dependent observations, r is an m-element error vector, and x is a p-element vector of regression coefficients to be found. The symbol $\leftarrow$ indicates 'is assumed to come from', so that $r \leftarrow N(0, \sigma^2 W)$ implies that the noise or error vector r is assumed to come from a normal distribution with mean $\mu = 0$ and covariance matrix $\sigma^2 W$, with W a known $m \times m$ symmetric non-negative definite matrix. We consider the use of QR factorization and singular-value decomposition (SVD) to solve (5.1) when W is (i) the identity, (ii) a nonsingular matrix, or (iii) a singular matrix. We summarize briefly each of these cases in turn.

(i) The case $W = I_m$ is associated with the *linear least-squares problem*:

$$\underset{x}{\text{minimize }} r^T r, \qquad r = b - Ax; \tag{5.2}$$

the solution discussed in Section 7.4, using QR factorization and SVD, depicts clearly the structure and sensitivity of (5.1).

(ii) The case when W is a nonsingular matrix is usually associated with the *weighted least-squares problem*:

$$\underset{x}{\text{minimize }} r^T W^{-1} r, \qquad r = b - Ax. \tag{5.3}$$

We can express (5.3) as a linear least-squares problem as follows. Let $W = FF^T$, for some matrix F, and put $s = F^{-1} r$, so that $s \leftarrow N(0, \sigma^2 I)$ and (5.3) becomes:

$$\underset{x}{\text{minimize }} s^T s, \qquad s = F^{-1} b - F^{-1} Ax, \tag{5.4}$$

which is in the form (5.2), and so can be solved as before, using QR factorization and SVD. However, this approach is numerically unstable when W is ill-conditioned; if W is singular, then (5.3) and (5.4) are not even defined. In these circumstances we proceed as follows.

(iii) Let s be any k-element vector such that $Fs = r$ and, as before, $W = FF^T$, for some $m \times k$ matrix F $(k \leq m)$. Rearranging (5.4) gives:

$$\operatorname*{minimize}_{x} s^T s, \qquad b = Ax + Fs. \qquad (5.5)$$

Although no more than a trivial rearrangement of (5.4), eqn (5.5) does not require the nonsingularity of F and hence W, and we still preserve the condition $r \leftarrow N(0, \sigma^2 W)$ because $s \leftarrow N(0, \sigma^2 I)$. Equation (5.5) is known as the *generalized linear least-squares problem*, and has been shown (Kourouklis and Paige 1981) to give the minimum-variance linear unbiased estimate for x in all cases of the general Gauss–Markov linear model.

We next address the question of solving (5.5).

7.6 The generalized linear least-squares problem

In this section we consider the generalized linear least-squares problem which, as we saw in Section 7.5, also arises in the context of solving the general Gauss–Markov linear model. We are required from (5.5) to solve the problem

$$\operatorname*{minimize}_{x} s^T s \text{ such that } b = Ax + Fs, \qquad (6.1)$$

where A and F are respectively $m \times p$ and $m \times k$ matrices, and where b, x, and s are respectively vectors of m, p, and k elements. We will again utilize the parallel algorithms developed for QR factorization in Chapter 6 and for singular-value decomposition (SVD) in Section 7.3.

Step 1 Perform a QR factorization of A, namely, $A = Q_A \begin{bmatrix} R_A \\ 0 \end{bmatrix}$, and substitute into (6.1) to give:

$$\operatorname*{minimize}_{x} s^T s \text{ such that } z = \begin{bmatrix} R_A \\ 0 \end{bmatrix} x + \begin{bmatrix} Q_1^T F \\ Q_2^T F \end{bmatrix} s, \qquad (6.2)$$

where $z = Q_A^T b$ and $Q_A = [Q_1, Q_2]$. If A has full column rank, then, for any s, we can find x as the solution of the upper triangular system of equations:

$$R_A x = z_1 - Q_1^T Fs, \qquad (6.3)$$

where $z = \begin{bmatrix} z_1 \\ z_2 \end{bmatrix}$, and (6.2) becomes:

$$\text{minimize}_{s}\ s^\mathsf{T}s \text{ such that } z_2 = Q_2^\mathsf{T}Fs. \qquad (6.4)$$

If A does not have full column rank, then we proceed to compute the SVD of R_A.

Step 2 Perform a QR factorization of $F^\mathsf{T}Q_2$, namely, $F^\mathsf{T}Q_2 = Q_F \begin{bmatrix} R_F \\ 0 \end{bmatrix}$, and substitute into (6.4) to give:

$$\text{minimize}_{s}\ s^\mathsf{T}s \text{ such that } z_2 = \begin{bmatrix} R_F \\ 0 \end{bmatrix}^\mathsf{T} Q_F^\mathsf{T}s,$$

which can be written as:

$$\text{minimize}_{p}\ p^\mathsf{T}p \text{ such that } z_2 = R_F^\mathsf{T}p_1, \text{ where } Q_F^\mathsf{T}s = p = \begin{bmatrix} p_1 \\ p_2 \end{bmatrix}. \quad (6.5)$$

Again, if R_F does not have full rank, then we proceed to compute the SVD of R_F; otherwise, we solve the lower triangular system of equations $R_F^\mathsf{T}p_1 = z_2$ for p_1, and taking $p_2 = 0$, we obtain

$$s^\mathsf{T}s = p_1^\mathsf{T}p_1, \quad \text{with } s = Q_F \begin{bmatrix} p_1 \\ 0 \end{bmatrix}. \qquad (6.6)$$

Hence (6.2) and (6.6) provide x and the residual vector s to satisfy (6.1).

The above approach has been described in detail by Paige (1978, 1979) and Hammarling (1987). So far, we have concentrated on the use of QR factorization and SVD: basic tools for which we have already discussed parallel algorithms. We now proceed to discuss the generalized singular-value decomposition, which provides a neat formulation of some of the preceding ideas.

7.7 Generalized singular-value decomposition

In this section, we generalize the singular-value decomposition (SVD) to any two matrices A ($m \times n$) and B ($p \times n$). The generalization is an important one, providing a numerically sound mathematical approach to a variety of problems, including the Kalman filter, least squares, and multivariate statistical analysis. The underlying concepts are not generally new, but it was not until 1976 that they were partially formalized by Van Loan. Paige and Saunders (1981) subsequently gave a complete general formulation which is numerically stable. Just as QR factorization and SVD are convenient techniques for solving the linear least-squares problem, generalized singular-value decomposition (GSVD) likewise performs a valuable function in the solution of the constrained and weighted least-squares problem. The detailed analysis of GSVD and proof of its existence are beyond the scope of this book. We therefore supplement our brief discussion with relevant references at the end of this section.

The GSVD for the matrix pair (A, B) may be defined as follows.

$$A = Q\begin{bmatrix} R \\ 0 \end{bmatrix}\begin{bmatrix} 0 & 0 & 0 \\ 0 & S & 0 \\ 0 & 0 & I \end{bmatrix}V^{\mathrm{T}}, \qquad B = Q\begin{bmatrix} R \\ 0 \end{bmatrix}\begin{bmatrix} I & 0 & 0 \\ 0 & C & 0 \\ 0 & 0 & 0 \end{bmatrix}U^{\mathrm{T}},$$

$$(7.1)$$

where Q, U, and V are orthogonal, R is a nonsingular upper triangular matrix, and C and S are diagonal matrices with

$$C^2 + S^2 = I, \qquad c_i > 0, \qquad s_i > 0.$$

The diagonal elements of C can be chosen in descending order, in which case the elements of S will be ascending. The pairs (c_i, s_i) are called the *generalized singular values* of (A, B). The matrices are assumed to be of compatible sizes.

If B is square and nonsingular then, clearly, from the collapse of certain diagonal blocks and the zero block of the R matrix, we have

$$B^{-1}A = U\begin{bmatrix} 0 & 0 \\ 0 & C^{-1}S \end{bmatrix}V^{\mathrm{T}}, \qquad (7.2)$$

which is the SVD of $B^{-1}A$. Further,

$$AA^{\mathrm{T}} = QR\begin{bmatrix} 0 & 0 \\ 0 & S^2 \end{bmatrix}(QR)^{\mathrm{T}}, \qquad BB^{\mathrm{T}} = QR\begin{bmatrix} I & 0 \\ 0 & C^2 \end{bmatrix}(QR)^{\mathrm{T}}, \quad (7.3)$$

thus QR is the congruence matrix that diagonalizes both AA^T and BB^T. The values s_i^2/c_i^2 are the nonzero eigenvalues of the generalized symmetric eigenvalue problem

$$AA^Tx = \lambda BB^Tx. \tag{7.4}$$

We now consider the application of GSVD (and SVD) to the matrix pair (A, F) in the generalized linear least-squares problem, given by (6.1), to highlight the structure and the sensitivity of the problem. We wish to find an estimate for x:

$$\underset{x}{\text{minimize }} s^Ts \text{ such that } b = Ax + Fs.$$

We first take the special case when F is a *nonsingular matrix* so that the problem becomes:

$$\underset{x}{\text{minimize }} s^Ts \text{ such that } F^{-1}b = F^{-1}Ax + s. \tag{7.5}$$

Substituting the SVD of $F^{-1}A$, from (7.2), leads to:

$$\underset{x}{\text{minimize }} s^Ts \text{ such that } s = F^{-1}b - U\begin{bmatrix} 0 & 0 \\ 0 & C^{-1}S \end{bmatrix}V^Tx. \tag{7.6}$$

Let

$$g = U^Ts, \quad U^TF^{-1}b = z = \begin{bmatrix} z_1 \\ z_2 \end{bmatrix}, \quad V^Tx = w = \begin{bmatrix} w_1 \\ w_2 \end{bmatrix},$$

$$D = C^{-1}S.$$

Then (7.6) may be reformulated as:

$$\underset{g}{\text{minimize }} g^Tg \text{ such that } g = z - \begin{bmatrix} 0 & 0 \\ 0 & D \end{bmatrix}w. \tag{7.7}$$

Thus we can solve for w_2 from $Dw_2 = z_2$ and, with w_1 arbitrary, we may find $x = Vw$; the residual vector is

$$s = Ug = U\begin{bmatrix} z_1 \\ 0 \end{bmatrix}.$$

In the more general case when F *can be a singular matrix*, we apply GSVD directly to (6.1) as follows. Substitute the GSVD of

A and *F* from (7.1) into (6.1):

$$b = Q \begin{bmatrix} R \\ 0 \end{bmatrix} \begin{bmatrix} 0 & 0 & 0 \\ 0 & S & 0 \\ 0 & 0 & I \end{bmatrix} V^{\mathsf{T}} x + Q \begin{bmatrix} R \\ 0 \end{bmatrix} \begin{bmatrix} I & 0 & 0 \\ 0 & C & 0 \\ 0 & 0 & 0 \end{bmatrix} U^{\mathsf{T}} s. \quad (7.8)$$

Hence

$$\begin{bmatrix} Q_1^{\mathsf{T}} b \\ Q_2^{\mathsf{T}} b \end{bmatrix} = \begin{bmatrix} R \\ 0 \end{bmatrix} \left\{ \begin{bmatrix} 0 & 0 & 0 \\ 0 & S & 0 \\ 0 & 0 & I \end{bmatrix} \begin{bmatrix} \beta_1 \\ \beta_2 \\ \beta_3 \end{bmatrix} + \begin{bmatrix} I & 0 & 0 \\ 0 & C & 0 \\ 0 & 0 & 0 \end{bmatrix} \begin{bmatrix} \mu_1 \\ \mu_2 \\ \mu_3 \end{bmatrix} \right\},$$

where

$$Q^{\mathsf{T}} = \begin{bmatrix} Q_1^{\mathsf{T}} \\ Q_2^{\mathsf{T}} \end{bmatrix}, \qquad \beta_i = V_i^{\mathsf{T}} x, \qquad \mu_i = U_i^{\mathsf{T}} s,$$

for $1 \le i \le 3$ with $V = [V_1, V_2, V_3]$ and $U^{\mathsf{T}} = [U_1, U_2, U_3]$. It follows that $Q_2^{\mathsf{T}} b = 0$, and we have

$$Q_1^{\mathsf{T}} b = R \begin{bmatrix} \mu_1 \\ S\beta_2 + C\mu_2 \\ \beta_3 \end{bmatrix}, \qquad \begin{bmatrix} \mu_1 \\ \mu_2 \end{bmatrix} \leftarrow N(0, \sigma^2 I), \quad (7.9)$$

which may be written as

$$Q_1^{\mathsf{T}} b = \begin{bmatrix} b_1 \\ b_2 \\ b_3 \end{bmatrix} = \begin{bmatrix} R_1 & R_{12} & R_{13} \\ 0 & R_2 & R_{23} \\ 0 & 0 & R_3 \end{bmatrix} \begin{bmatrix} \mu_1 \\ S\beta_2 + C\mu_2 \\ \beta_3 \end{bmatrix}. \quad (7.10)$$

We note that *R* is upper triangular and nonsingular with positive diagonal elements.

Observations from (7.10)

(i)   The vector β_1 has no effect on *b*, but clearly β_2 and β_3 do affect *b*.
Thus

$$x = x^{\mathrm{n}} + x^{\mathrm{e}}, \qquad x^{\mathrm{n}} = V_1 \beta_1, \qquad x^{\mathrm{n}} = V_2 \beta_2 + V_3 \beta_3,$$

where x^{e} refers to the *estimable* part of *x* and x^{n} to the *nonestimable* part of *x*.

(ii)   We can determine β_3 exactly from

$$b_3 = R_3 \beta_3. \quad (7.11)$$

This possibility arises only if F does not have full rank, that is, if W in (5.1) is singular.

(iii) We can estimate $\boldsymbol{\beta}_2$ from

$$R_2(S\boldsymbol{\beta}_2 + C\boldsymbol{\mu}_2) + R_{23}\boldsymbol{\beta}_3 = \boldsymbol{b}_2, \qquad \boldsymbol{\mu}_2 \leftarrow N(\boldsymbol{0}, \sigma^2 I).$$

Thus $R_2^{-1}(\boldsymbol{b}_2 - R_{23}\boldsymbol{\beta}_3) = S\boldsymbol{\beta}_2 + C\boldsymbol{\mu}_2$, and the left-hand side of this equation is known once $\boldsymbol{b}$ is observed.

(iv) The vector $\boldsymbol{\mu}_3$ has no effect on $\boldsymbol{b}$.

(v) The random vector $\boldsymbol{\mu}_1$ can be determined exactly by solving (7.10) once $\boldsymbol{b}$ is observed. Since $\boldsymbol{\mu}_1 \leftarrow N(\boldsymbol{0}, \sigma^2 I)$, this represents the vector of uncorrelated regression residuals.

Having estimated $\boldsymbol{\beta}_i$ and $\boldsymbol{\mu}_i$, we can recover x and the residual vector s from $\boldsymbol{\beta}_i = V_i^T x$ and $\boldsymbol{\mu}_i = U_i^T s$.

The transformed model (7.10) thus reveals the sensitivity of the observed $\boldsymbol{b}_i$ to the unknown $\boldsymbol{\beta}_i$ and random $\boldsymbol{\mu}_i$, particularly if R_1 and R_3 are arranged to be diagonal matrices. If, for example, R_3 has a few small singular values, then from (7.11) we can see that the corresponding values of $\boldsymbol{\beta}_3$ will not affect $\boldsymbol{b}_3$ strongly, and it will be difficult to compute these elements accurately in the presence of rounding errors. Usually $\boldsymbol{\beta}_3$ will affect $\boldsymbol{b}_2$ and $\boldsymbol{b}_1$, but only $\boldsymbol{b}_3$ is used in computing $\boldsymbol{\beta}_3$.

Similarly, if R_1 has some small singular values, then from (7.10) we observe that the corresponding values of $\boldsymbol{\mu}_1$ will not affect $\boldsymbol{b}$ strongly, and again it will be difficult to compute these elements accurately in the presence of rounding errors. This argument also applies to the computation of $d_2 = S\boldsymbol{\beta}_2 + C\boldsymbol{\mu}_2$. A further source of error is when S has very small diagonal elements, which will result in C having elements close to unity, and thus the effect of some elements of $\boldsymbol{\beta}_2$ on d_2 will tend to be small compared with some elements of $\boldsymbol{\mu}_2$. For further details see Paige (1985b).

For proof of the existence of GSVD and a discussion on related properties, the reader is referred to earlier references in this section. The Kogbetliantz approach applied to GSVD is held to be suitable for systolic implementation (Paige 1986).

7.8 Singular-value decomposition for a matrix product

In this section, we consider the computation of certain con-tragredient transformations which are particularly useful in linear system theory. It transpires, as we will see later, that the problem is actually one of computing the eigenvectors of a generalized eigenvalue problem (Chapter 5, Section 5.12). However, the evaluation of the coefficient matrices—or the matrix pair (A, B)—squares the condition number and the eigenvalue approach is therefore numerically unstable (the situation is analogous to that encountered in solving the linear least-squares problem through the use of normal equations, see Section 7.4).

An alternative approach is to use the generalized singular value decomposition (GSVD) introduced in Section 7.7. Unfortunately it turns out that the method is not equivalent to the type of eigenproblem needed to evaluate contragredient transformations. We therefore describe an SVD of a product of two nondefinite matrices proposed by Fernando and Hammarling (1988) and referred to as PSVD. The definition is a variant of the SVD of a product suggested by Heath *et al.* (1986), and is chosen so that it is equivalent to the type of eigenproblem required for the computation of the contragredient transformation.

We will discuss these issues in detail and indicate particular areas of application. In computing PSVD, we rely once again on the parallel algorithms developed earlier: the Kogbetliantz method (essentially the Jacobi method for asymmetric matrices), QR factorization, and SVD.

First we define a contragredient transformation: Let A, B, $\bar{A}$, and $\bar{B}$ be $m \times n$ matrices and let X be an $n \times n$ nonsingular matrix. If

$$\bar{A} = AX \quad \text{and} \quad \bar{B} = BX^{-\mathsf{T}},$$

then $(A, B) \mapsto (AX, BX^{-\mathsf{T}})$ is called a *contragredient* transformation. (Here $X^{-\mathsf{T}}$ denotes $(X^{-1})^{\mathsf{T}}$.)

We are particularly interested in finding the matrix X of a contragredient transformation

$$(A, B) \mapsto (\bar{A}, \bar{B}) = (AX, BX^{-\mathsf{T}}) \tag{8.1}$$

such that $\bar{A}^{\mathsf{T}}\bar{A}$ and $\bar{B}^{\mathsf{T}}\bar{B}$ are diagonal. Theoretically we can find X as follows. We have

$$(X^{\mathsf{T}}A^{\mathsf{T}}AX)(X^{-1}B^{\mathsf{T}}BX^{-\mathsf{T}}) = \Sigma^2 \Delta^2, \tag{8.2}$$

where the diagonal matrices Σ and Δ are such that

$$\Sigma^2 = \bar{A}^{\mathsf{T}}\bar{A}, \qquad \Delta^2 = \bar{B}^{\mathsf{T}}\bar{B}.$$

Hence the columns of X are eigenvectors of the problem

$$(A^{\mathrm{T}}A)(B^{\mathrm{T}}B)x = \lambda x, \tag{8.3}$$

and, provided that the eigenvalues given by the elements of the diagonal matrix $\Sigma^2 \Delta^2$ in (8.2) are distinct, then X is unique up to permutation and scalar multiplication of the columns.

We encounter yet again the problem that, in forming $A^{\mathrm{T}}A$ and $B^{\mathrm{T}}B$ in (8.3), the data are squared three times, and this can lead to serious loss of accuracy. Comparing (8.3) with (7.4), we see that, even with the GSVD approach, we have to invert $B^{\mathrm{T}}B$ or $A^{\mathrm{T}}A$ in order to relate it to the eigenproblem (8.3), and this inversion can be numerically unstable, particularly if $B^{\mathrm{T}}B$ (or $A^{\mathrm{T}}A$) is ill conditioned. To overcome these difficulties, we consider a variant of SVD for the matrix product AB^{T}.

Let A and B be $m \times n$ and $p \times n$ matrices so that

$$A = U\Sigma R Q^{\mathrm{T}}, \qquad B = V\Delta R^{-\mathrm{T}}Q^{\mathrm{T}}, \qquad AB^{\mathrm{T}} = U\Sigma\Delta V^{\mathrm{T}}, \tag{8.4}$$

where U and V have orthonormal columns, Q is orthogonal, R is an $n \times n$ upper triangular matrix with full column rank, and Σ and Δ are diagonal matrices with non-negative elements. The matrices U and V are $m \times n$ and $p \times n$ respectively, and R, Q, Σ, and Δ are $n \times n$. The singular values of the matrix product AB^{T} are given by the elements of the diagonal matrix $\Sigma\Delta$. The decomposition in (8.4) is referred to as PSVD. From (8.4) we obtain

$$(A^{\mathrm{T}}A)(B^{\mathrm{T}}B) = (QR^{\mathrm{T}})(\Sigma^2\Delta^2)(QR^{\mathrm{T}})^{-1}, \tag{8.5}$$

which is the required form of the eigenvalue problem given in (8.3). The contragredient transformation is thus given by

$$X = QR^{-1} \qquad \text{and} \qquad X^{-1} = RQ^{\mathrm{T}}. \tag{8.6}$$

There is, of course, no need to compute the inverse of X; we can simply accumulate the product RQ^{T} during the decomposition.

A more rigorous definition of PSVD, incorporating all null-spaces, together with a proof of its existence, is given by Fernando and Hammarling (1988). The computational steps are also described by the authors and are based on the Kogbetliantz algorithm for SVD (Section 7.3). The main difference is that we compute the QR decomposition of matrix A as for the SVD, but the QL decomposition of matrix B when the pair (A, B) is being considered. Fernando and Hammarling also indicate the use of PSVD to compute balanced realizations of linear systems, and to solve Lyapunov equations. For further applications in circuit design, signal processing, and control, the reader may also like to consult Newcomb (1960), Mullis and Roberts (1976), and Laub *et al.* (1987), respectively.

7.9 The Kalman filtering problem

In this section, we consider the Kalman filtering problem and show how it can be expressed as a least-squares problem. It is then possible to approach it via the parallel methods for singular-value decomposition (SVD), QR factorization, and least-squares analysis discussed in earlier sections. The problem was named after Kalman in 1960 and is of crucial importance in signal processing and related areas.

We consider a simple form of the problem posed as follows (Hammarling 1986).

Let

$$x_k = A_k x_{k-1} + u_k, \qquad \mathsf{E} u_k = 0, \qquad \mathsf{E}(u_k u_k^\mathsf{T}) = U_k,$$
$$y_k = C_k x_k + v_k, \qquad \mathsf{E} v_k = 0, \qquad \mathsf{E}(v_k v_k^\mathsf{T}) = V_k, \qquad (9.1)$$

where

x_k is the n-vector of state variables,
y_k is the m-vector of observed (measured) variables,
u_k is the n-vector of process noise,
v_k is the m-vector of observation (measurement) noise.

The notations $\mathsf{E} u_k$ and cov $u_k = \mathsf{E}(u_k u_k^\mathsf{T})$ indicate the mean and the covariance of u_k respectively. It is usually assumed that u_k and v_k come from a *normal distribution*, but this is not strictly necessary. The problem we consider is a particular case of the Wiener filtering problem: to find the minimum-variance linear unbiased estimate $\hat{x}_k$ of x_k, for given observations $y_0, \ldots, y_k$, in the sense that tr cov$(x_k - \hat{x}_k \mid y_0, \ldots, y_k)$ is minimized.

Kalman gave the following solution. Let

$$\mathsf{E}(x_k x_k^\mathsf{T}) = X_k, \qquad \mathsf{E}(y_k y_k^\mathsf{T}) = Y_k, \qquad \mathsf{E}(\hat{x}_k \hat{x}_k^\mathsf{T}) = \hat{X}_k.$$

Substituting from (9.1) gives

$$X_k = A_k X_{k-1} A_k^\mathsf{T} + U_k, \qquad Y_k = C_k X_k C_k^\mathsf{T} + V_k.$$

We now define

$$\bar{x}_k = A_k \bar{x}_{k-1}, \qquad \bar{X}_k = A_k \hat{X}_{k-1} A_k^\mathsf{T} + U_k, \qquad (9.2)$$

$$\bar{Y}_k = C_k \bar{X}_k C_k^\mathsf{T} + V_k, \qquad K_k = \bar{X}_k C_k^\mathsf{T} \bar{Y}_k^{-1}, \qquad (9.3)$$

where

$\bar{x}_k$ is referred to as the *predictor* or *estimator*,
$\bar{X}_k$ is the *predicted covariance matrix*,
K_k is the *Kalman gain matrix*.

Equation (9.2) is called the *time update*. Kalman showed that

$$\hat{X}_k = \bar{X}_k - \bar{X}_k C_k^\mathsf{T} \bar{Y}_k^{-1} C_k \bar{X}_k \qquad (9.4)$$

and

$$\hat{x}_k = \bar{x}_k + K_k(y_k - C_k \bar{x}_k). \qquad (9.5)$$

These are referred to as the *measurement update*; eqn (9.5) has also become known as the *Kalman filter*. The calculation of the measurement update directly from eqns (9.4) and (9.5) can lead to serious numerical difficulties, particularly when computing the products of the form $B^\mathsf{T} A B$ in (9.4), where valuable information can be lost when the matrices are nearly orthogonal or nearly rank-deficient. The computation of $\bar{Y}_k^{-1}$ in (9.4) can also give rise to numerical inaccuracy, particularly when $\bar{Y}_k$ is ill conditioned or nearly singular. To avoid these difficulties, and to produce a numerically reliable solution for computing the Kalman filter, we take a least-squares approach as follows.

Suppose we have an initial estimate $\hat{x}_0$ so that

$$x_0 = \hat{x}_0 + u_0. \qquad (9.6)$$

Then, for $k = 0, \ldots, k$, eqn (9.1) gives

$$x_0 = \hat{x}_0 + u_0, \qquad y_0 = C_0 x_0 + v_0,$$

$$x_1 = A_1 x_0 + u_1, \qquad y_1 = C_1 x_1 + v_1,$$

$$\vdots \qquad\qquad \vdots$$

$$x_k = A_k x_{k-1} + u_k, \qquad y_k = C_k x_k + v_k,$$

which can be written as

$$b_k = B_k z_k + e_k, \tag{9.7}$$

with

$$b_k = (-\hat{x}_0, y_0, 0, y_1, 0, \dots, y_{k-1}, 0, y_k),$$

$$z_k = (x_1, \dots, x_k),$$

$$e_k = (u_0, v_0, u_1, v_1, \dots, u_k, v_k),$$

where the parenthetic listing of vectors $(a, b, \dots)$ indicates the stacked column vector $[a^\mathsf{T}, b^\mathsf{T}, \dots]^\mathsf{T}$ and the zero vectors are of dimension n, and where

$$B_k = \begin{bmatrix}
-I & 0 & 0 & \cdots & 0 & 0 \\
C_0 & 0 & 0 & \cdots & 0 & 0 \\
A_1 & -I & 0 & \cdots & 0 & 0 \\
0 & C_1 & 0 & \cdots & 0 & 0 \\
0 & A_2 & -I & \cdots & 0 & 0 \\
\cdot & \cdot & \cdot & & \cdot & \cdot \\
0 & 0 & 0 & \cdots & A_k & -I \\
0 & 0 & 0 & \cdots & 0 & C_k
\end{bmatrix}.$$

Assuming that $\mathsf{E}(u_i u_j^\mathsf{T}) = \mathsf{E}(v_i v_j^\mathsf{T}) = 0$ ($i \neq j$) and $\mathsf{E}(u_i v_j^\mathsf{T}) = 0$, for all i and j, and putting $W_k = \mathsf{E}(e_k e_k^\mathsf{T})$, then, from (9.6) and (9.7), we obtain the block diagonal matrix

$$W_k = \mathrm{diag}(U_0, V_0, \dots, U_k, V_k).$$

It can be shown that solving the Kalman filtering problem is equivalent to determining the component $\hat{x}_k$ of the solution $\hat{z}_k$ to the *weighted* least-squares problem (Duncan and Horn 1972):

$$\text{minimize } e_k^\mathsf{T} W^{-1} e_k \text{ such that } b_k = B_k \hat{z}_k + e_k. \tag{9.8}$$

We can express (9.8) as a *linear* least-squares problem as follows.

Let R_k denote the Choleski factor of W_k, i.e. $W_k = R_k^\mathsf{T} R_k$, where

$$R_k = \begin{bmatrix}
G_0 & 0 & \cdots & 0 & 0 \\
0 & H_0 & \cdots & 0 & 0 \\
\vdots & & \ddots & & \vdots \\
0 & 0 & \cdots & G_k & 0 \\
0 & 0 & \cdots & 0 & H_k
\end{bmatrix}$$

and G_k and H_k are the Choleski factors of U_k and V_k respectively. If we now let $r_k = R_k^{-\mathsf{T}} e_k$, so that $E(r_k r_k^{\mathsf{T}}) = I$, then (9.8) becomes

$$\text{minimize } r_k^{\mathsf{T}} r_k \text{ such that } R_k^{-\mathsf{T}} b_k = R_k^{-\mathsf{T}} B_k \hat{z}_k + r_k. \qquad (9.9)$$

The computation of (9.8) and (9.9) is described in Sections 7.5 and 7.6. The parallel algorithms developed for QR factorization in Chapter 6, and for SVD and GSVD in Sections 7.3 and 7.7 of this chapter, again form the basis of the computation.

For a detailed description the reader is referred to the original Kalman (1960) paper. Although the above analysis relates to the filtering problem, the techniques discussed can be extended to allow smoothing and prediction. For a complete generalization, and a unification of Kalman filtering and information filtering, see Paige (1985a).

7.10 Eigensolutions via singular-value decomposition

From the definition of the SVD (eqn (2.2) of Section 7.2), we have $\sigma(A) = \sqrt{\text{ev}}\, A^{\mathsf{T}} A$, where $\sigma(A)$ indicates the singular values of A and ev $A^{\mathsf{T}} A$ indicates the eigenvalues of $A^{\mathsf{T}} A$. If A is a symmetric matrix of order n, then $A^{\mathsf{T}} A = A^2$ and $\sigma(A) = \sqrt{\text{ev}\, A^2} = \sqrt{(\text{ev}\, A)^2} = |\text{ev}\, A|$, i.e.

$$\sigma(A) = |\text{ev}\, A|. \qquad (10.1)$$

Thus we can compute the eigenvalues via the SVD. Applying the method of Section 7.3, we have $B = U \begin{bmatrix} \Sigma \\ 0 \end{bmatrix} = A P^{(1)} \cdots P^{(s)}$, where the $P^{(k)}$ are plane rotations and s is the final iteration step. Failure may occur, however, if $B = [b_1, \ldots, b_n]$ is singular. The presence of a null column b_i in B gives a zero singular value, and the identity $A = U \begin{bmatrix} \Sigma \\ 0 \end{bmatrix} V^{\mathsf{T}}$ is valid. However, a null b_i only gives a trivial solution for the eigenproblem $Ax = \lambda x$. We can circumvent this problem by considering, instead of A, the positive definite matrix

$$\bar{A} = A + cI, \qquad (10.2)$$

for some sufficiently large $c > 0$. Now $\bar{A}$ has the same eigenvectors as A, and the eigenvalues are related by $\lambda_i = \bar{\sigma}_i - c$ ($i = 1, \ldots, n$), where $\bar{\sigma}_i$ are the singular values of $\bar{A}$.

1.6343832138O1O99	0.3O4938846O32O23	3.2719648O585460	0.2513392158490288
0.8523697594022732	0.399O63515288680l	5.91825937823385O	0.2865703642814O34
0.9212214243458881	0.45O70576577432 42	1O.O6O591788O4632	0.257408462255657 1
1.3184471651888858	0.25O148418452O392	0.546528O266906547	0.255414715198514O
20.8994081671683l	0.2597348347454899	0.738421991511258 5	0.691006303173274 9
67.5268278295562 2	0.3594265712962532	3.9O7758399132313	0.518953221521980 2
2.3980411468O5O33	0.253743761444796 7	1.46284669706358 7	0.328743992573903 6
0.6487601605716985	1.19576864987 3666	1686.169152796724	0.2688355612414388
0.320126629O453243	0.34832274188373 25	0.79186265961750 39	0.276832779593868 5
2.09O31176822 5244	0.2915867703 68322	4.754830906719853	0.298269204464125 7
0.371517550845687 9	0.25238735186 5O27	7.577607O55233296	0.3120909624713 97
0.431914311280808 8	0.3846929496104334	14.O18226731884 71	0.9999999999999 499
1.84O2747435OO968	0.6109757O265 52269	0.41475696533616 4	0.272628459597268 3
34.4933691947424O	0.262405448799895 O	0.49398503333282517	0.338120412981492 6
187.42624124133 5865	0.281471452599596 5	2.782638162O6954O	0.2654338298460694
1.O906831773 52352	0.471321957577721 3	0.57706316O1O5740O	0.25O594O34251652 6

Table 10.1   Eigenvalues of Frank Matrix of order 64 using one-sided SVD Algorithm[1]

[1] The results are consistent with the analytical solution

If A is known to be positive or negative semidefinite, then the eigenvalues may be found by taking the square root with the appropriate sign. If A is not definite, the sign of the eigenvalues may be found via (Kaiser 1972):

$$\lambda_i^3 = \boldsymbol{u}_i A \boldsymbol{u}_i^{\mathsf{T}},$$

where $\boldsymbol{u}_i$ is the ith column of U.

The shift strategy afforded by (10.2) is well suited to parallel implementation. Table 10.1 gives, in the case $n = 64$, the results obtained for the Frank matrix

$$F_n = \begin{bmatrix} n & n-1 & n-2 & \cdots & 1 \\ n-1 & n-1 & n-2 & \cdots & 1 \\ n-2 & n-2 & n-2 & \cdots & 1 \\ \vdots & \vdots & \vdots & & \vdots \\ 1 & 1 & 1 & \cdots & 1 \end{bmatrix},$$

which is known to have the eigenvalues $\dfrac{1}{2}\left(1 - \cos\dfrac{(2i-1)\pi}{2n+1}\right)^{-1}$ $(i = 1, \ldots, n)$.

By using the SVD to compute the eigensolution, only one-sided transformations are required. This halves the number of parallel steps involved in the multiplication in comparison with the parallel Jacobi method, and consequently there is a saving in both time and storage.

Postscript

Algorithms were constructed in the past with a sequential model of computation in mind, which facilitated their execution on serial machines. With the recent development of parallel machines, this is no longer appropriate. Parallel computing requires us to construct algorithms that can be executed on independent processors simultaneously. Because of the fundamental difference between serial and parallel computers, the algorithm design for one is generally incompatible with the other. This does not imply, however, that serial techniques are irrelevant to the development of parallel algorithms.

In pragmatic terms, the design of algorithms is strongly influenced by prevailing architectures: different systems automatically impose different priorities. With SIMD systems and vector processors, the primary aim is to divide the problem into small identical parts which require identical instructions to act on independent sets of data. An SIMD system requires that data be placed in separate processing elements; a vector processor, on the other hand, requires that they be placed in the separate indices of a set of vectors so that a vector instruction may be used. Many MIMD systems contain relatively few processors, working asynchronously (which may be allocated to entirely separate problems), and it is particularly important to ensure that use of these processors is maximized. In many SIMD systems, however, where the available processors are comparatively numerous, with the majority in operation most of the time, speed of processing is more crucial than full utilization of processors. In practice, of course, an architecture is often a hybrid form, incorporating elements of some or all of the above systems. The way in which a given system is exploited is also controlled by the intellectual trends of the day, with each generation interpreting received ideas from a different perspective.

While serial and parallel programming are largely incompatible, a comparison of the two is nonetheless instructive and, throughout this book, one has been a constant point of reference for the other. Some of the key areas of divergence are as follows.

Firstly, parallel computing depends on an efficient arrangement of data in the memory—something which is of no particular significance on a serial machine. Secondly, serial algorithms that appear to be intrinsically serial may harbour a hidden parallelism. Thirdly, the numerical properties of parallel algorithms may differ substantially from those of their serial counterparts.

Thus, in seeking to promote parallelism, we should not discard serial techniques, but begin by re-examining them. The performance of many of the best sequential algorithms on parallel machines is disappointing. Conversely, some of the worst may surprise us: an inefficient sequential algorithm may have the makings of an extremely efficient parallel one, as with the Jacobi method for the eigenvalue problem. To increase the overall speed of computation, therefore, we should seek not only to devise new ways of constructing parallel algorithms, but also, by 'thinking in parallel', to exploit the latent parallelism in older techniques.

Bibliography

Besides the publications referred to in the text, a number of others are included.

Adams, L. (1982). Iterative algorithms for large sparse linear systems on parallel computers. Ph.D. dissertation. University of Virginia.

Adams, L., and Ortega, J. (1982). A multi-color SOR method for parallel computation. *Proceedings of the international conference on parallel processing.*

Aho, A., Hopcroft, J., and Ullman, J. (1974). *The design and analysis of computer algorithms.* Addison Wesley.

Ajtai, M., Komlós, J., and Szemerédi, E. (1983). An $O(n \log n)$ sorting network. *Proceedings 15th. ACM symposium on theory of computing.* New York.

Akl, S. G. (1985). *Parallel sorting algorithms.* Academic Press.

Arnot, N. R., Wilkinson, G. G., and Burge, R. E. (1982). Application of the ICL-DAP for two-dimensional optical image processing. *Computer Physics Communications,* **26**, North Holland.

Ball, W. W. R., and Coxeter, H. S. (1939). *Mathematical recreations and essays.* Macmillan.

Barlow, J., and Ipsen, I. (1984). Research report Yale U DCS RR-310, Dept. of Computer Science, Yale University.

Barlow, R., Evans, D., and Shanehchi, J. (1984). Sparse matrix vector multiplication on the DAP. In *Vector and array processors* (ed. D. Paddon). Oxford University Press.

Batcher, K. E. (1968). Sorting networks and their applications. Proceedings AFIPS Spring JCC, **32**.

Bathé, K.-J., and Wilson, E. (1976). *Numerical methods in finite element analysis.* Prentice Hall, Englewood Cliffs, N.J.

Bentin, J., and Modi, J. J. (1988). The siftsort algorithm. Internal report CUED/F-INFENG/TR.9, Engineering Dept., Cambridge University.

Bently, J. L., and Kung, H. T. (1979). A tree machine for searching problems. In *Proceedings of the international conference on parallel processing.*

Biswas, R., and Amaratunga, G. (1987). A banded Gauss–Jordan solver. Internal report, Engineering Dept., Cambridge University.

Bitton, D., DeWitt, D., Hsiao, D., and Menon, J. (1984). A taxonomy of parallel sorting. *Computing Surveys,* **16** (3).

Bojanczyk, A., Brent, R. P., and Kung, H. T. (1984). Numerically stable solution of dense systems of linear equations using mesh-connected processors. *SIAM Journal on Scientific and Statistical Computing*, **5** (1).

Bollabás, B. (1979). *Graph theory*. Springer Verlag.

Bollabás, B., and Hell, P. (1985). Sorting and graphs. In *Graphs and orders* (ed. D. Reidel). D. Reidel Publishing Company.

Borodin, A.. and Munro, I. (1975). *The computational complexity of algebraic and numerical problems*. American Elsevier, New York.

Bowden, B. W. (1953). *Faster than thought*. Pitman.

Bowgen, G. S. J., and Modi, J. J. (1985). Implementation of QR factorization on the DAP using Householder transformations. *Computer Physics Communications*, **37**.

Boyle, C., and Parkinson, D. (1985). Sorting on a bit organised parallel computer. In *Parcella*-84 (eds. W. Handler, T. Legendi, and G. Wolf). Akademi Verlag.

Brent, R. P. (1973). The parallel evaluation of arithmetic expressions in logarithmic time. In *Complexity of sequential and parallel numerical algorithms* (ed. J. Traub). Academic Press.

Brent, R. P. (1974). The parallel evaluation of general arithmetic expressions. *Journal of the Association of Computer Machinery*, **21** (2).

Brent, R. P., and Luk, F. T. (1982). A systolic architecture for the singular value decomposition. Tech. Report TR 82-522, Dept. of Computer Science, Cornell University, Ithaca, New York, 14853.

Brent, R. P., and Luk, F. T. (1982). A systolic architecture for almost linear time solution of the symmetric eigenvalue problem. Tech. Report TR 82-525, Dept. of Computer Science, Cornell University, Ithaca, New York, 14853.

Brodlie, K. W. (1973). Ph.D. dissertation, Dept. of Mathematics, University of Dundee.

Brodlie, K. W., and Powell, M. J. D. (1975). On the convergence of cyclic Jacobi methods. *Journal of the Institute of Mathematics and its Applications*, **15**.

Buzbee, B. L. (1984). Plasma simulation and fusion calculation. In *High-speed computation* (ed. J. Kowalik). Springer-Verlag.

Charlier, J.-P., and Van Dooren, P. (1986). On Kogbetliantz's SVD algorithm in the presence of clusters. Manuscript M155, Philips Research Laboratory, Brussels.

Charlier, J.-P., Vanbegin, M., and Van Dooren, P. (1986). On efficient implementation of Kogbetliantz's algorithm for computing the singular value decomposition. Manuscript M154, Philips Research Laboratory, Brussels.

Chen, S. C. (1975). Speedup of iterative programs in multiprocessing systems. Dissertation, Dept. of Computer Science. University of Illinois, Urbana.

Cline, A., Moler, G., Stewart, G., and Wilkinson, J. (1979). An estimate for the condition number of a matrix. *SIAM Journal on Numerical Analysis*, **16**.

Cosnard, M., and Robert, Y. (1986). Parallel QR decomposition of a rectangular matrix. *Numerische Mathematik*, **48**.

Davies, R. O., and Modi, J. J. (1986*a*). A direct method for completing eigenproblem solutions on a parallel computer. *Linear Algebra and its Applications*, **77**.

Davies, R. O., and Modi, J. J. (1986*b*). Complete parallel sorting schemes. In *Parallel Computing*, **85** (eds. M. Feilmeier, G. Joubert, and U. Schendel), North Holland.

Dixon, L. C. W., and Duxbury, P. G. (1985). Finite element optimization on the DAP. *Computer Physics Communications*, **37**.

Dongarra, J. J., Sameh, A. H., and Sorensen, D. C. (1986). Implementation of some concurrent algorithms for matrix factorization. *Parallel Computing*, **3**.

Duff, I. S. (1986). Parallel implementation of multifrontal schemes. *Parallel Computing*, **3**.

Duff, I. S. (1987). The influence of vector and parallel processors on numerical analysis. In *The state-of-the-art in numerical analysis* (eds. A. Iserles and M. J. D. Powell). Oxford University Press.

Duff, I. S., and Reid, J. (1982). Experience of sparse matrix codes on the CRAY-1. *Computer Physics Communications*, **26**.

Duncan, D., and Horn, S. (1972). Linear dynamic recursive estimation from the viewpoint of regression analysis. *Journal of the American Statistical Association*, **67**.

Egbert, W. E. (1978). Personal calculation algorithms IV: logarithmic functions. *Hewlett-Packard Journal*, **29** (8).

Erickson, J. (1972). Iterative and direct methods for solving Poisson's equation and their adaptability to Illiac IV. Centre for advanced computation, No. 60, University of Illinois, Urbana.

Evans, D. J. (1982). Parallel numerical algorithms for linear systems. In *Parallel processing systems* (ed. D. Evans). Cambridge University Press.

Evans, D. J. (1984). Parallel S.O.R. iterative methods. *Parallel Computing*, **1** (1).

Evans, D. J., Hadjidimos, A., and Noutsos, D. (1981). The parallel solution of banded linear equations by the new quadrant interlocking factorization (Q.I.F.) method. *International Journal of Computational Mathematics*, **9**.

Even, S. (1974). Parallelism in tape-sorting. *Communications of the ACM*, **17** (4).

Feilmeier, M. (1982). Parallel numerical algorithms. In *Parallel processing systems* (ed. D. Evans). Cambridge University Press.

Fernando, K. V., and Hammarling, S. J. (1988). A product induced singular value decomposition (∏SVD) for two matrices and balanced realisation. In *Linear algebra in signals, systems and control* (ed. B. N. Datta). SIAM.

Feurzeig, W. (1960). A.C.M. algorithm 23: Mathsort. *Communication of the Association for Computer Machinery,* **3** (11).

Flanders, P. M. (1982). A unified approach to a class of data movements on an array processor. *IEEE Transactions on Computers,* **C-31** (9).

Flanders, P. M., and Parkinson, D. (1988). Musical Bits—a calculus of data organisation for effective parallel processing. *Future Computing,* **2** (2).

Fletcher, R. (1976). Conjugate gradient methods for indefinite systems. In *Proceedings of the Dundee conference on numerical analysis*—1975 (ed. G. A. Watson). Springer-Verlag.

Flynn, M. J. (1966). Very high speed computing systems. *Proceedings of the IEEE,* **54** (12).

Forsythe, G. E., and Henrici, P. (1960). The cyclic Jacobi method for computing the principal values of a complex matrix. *Transactions of the American Mathematical Society,* **94**.

Gannon, D. (1980). A note on pipelining a mesh connected multi-processor for finite element problems by nested dissection. *Proceedings of the international conference on parallel processing.* IEEE Computer Society.

Gannon, D., and Van Rosendale, J. (1982). Highly parallel multigrid solvers for elliptic PDEs: an experimental analysis. ICASE, NASA Langley, Hampton, Virginia.

Gentleman, W. M. (1973). Least squares computations by Givens transformations without square roots, *Journal of the IMA,* **12**.

Gentleman, W. M. (1975). Error analysis of QR decomposition by Givens transformations. *Linear algebra and its applications,* **10**.

Gentleman, W. M. (1978). Some complexity results for matrix computations on parallel processors. *Journal of the Association for Computer Machinery,* **25** (1).

Gentleman, W. M., and Kung, H. T. (1981). Matrix triangularization by systolic arrays. Proceedings SPIE Symposium, **298**, Real-time signal processing IV. The Society of Photo-optical Instrumentation Engineers.

George, A., Heath, M., and Liu, J. (1986). Parallel Cholesky factorization on a shared-memory multiprocessor. *Linear algebra and its applications,* **77**.

Ghouila-Houri (1960). Une condition suffisante d'existence d'un circuit hamiltonien. *Comptes rendus de l'Académie scientifique,* **251** (494), Paris.

Givens, W. (1958). Computation of plane unitary rotations transforming a general matrix to triangular form. *Journal of the Society for Industrial and Applied Mathematics,* **6**.

Golub, G. H., and Kahan, W. (1965). Calculating the singular values and pseudo-inverse of a matrix. *SIAM Journal on Numerical Analysis,* **2.**

Golub, G. H., and Van Loan, C. F. (1983). *Matrix computations.* North Oxford Academic, Oxford.

Gostick, R. (1979). Software and algorithms for the Distributed Array Processor. *ICL Technical Journal,* **1.**

Gregory, R. T. (1953). Computing eigenvalues and eigenvectors of a symmetric matrix on the ILLIAC. *Mathematical tables and other aids to computation,* **7.**

Gurd, J. R. (1982). Data flow architecture. Information Technology state-of-the-art report. In *Supercomputer technologies.* Pergamon.

Gurd, J. R. (1988). A taxonomy of parallel computer architectures. *Proceedings of the international conference on the design and application of parallel digital processors.* Lisbon, Portugal. IEE.

Hackbusch, W. (1978). On the multigrid method applied to difference equations. *Computing,* **20.**

Hammarling, S. J. (1974). A note on the modification to the Givens plane rotation. *Journal of the Institute of Mathematics and its Applications,* **13.**

Hammarling, S. J. (1986). The numerical solution of the Kalman filtering problem. In *Computational and combinatorial methods in systems theory* (eds. C. Byrnes, A. Lindquist). North Holland.

Hammarling, S. J. (1987). The numerical solution of the general Gauss-Markov linear model. In *Mathematics of signal processing* (eds. T. Durrani, J. Abbiss, J. Hudson, R. Madan, J. McWhirter, and T. Moore). Oxford University Press.

Harary, F. (1969). *Graph theory.* Addison Wesley.

Heath, M., Laub, A., Paige, C., and Ward, R. (1986). Computing the singular value decomposition of a product of two matrices. *SIAM Journal on Scientific and Statistical Computing,* **7** (4).

Heller, D. (1978). A survey of parallel algorithms in numerical linear algebra. *SIAM Review,* **20.**

Heller, D., and Ipsen, I. (1983). Systolic networks for orthogonal decomposition. *SIAM Journal on Scientific and Statistical Computing,* **4** (2).

Hellier, R. L. (1982). DAP implementation of the WZ algorithm. *Computer Physics Communications,* **26.**

Henrici, P. (1958). On the speed of convergence of cyclic and quasicyclic Jacobi methods for computing eigenvalues of hermitian matrices. *Journal of the Society for Industrial and Applied Mathematics,* **6.**

Hestenes, M. R. (1958). Inversion of matrices by biorthogonalization and related results. *Journal of the Society for Industrial and Applied Mathematics,* **6.**

Hoare, C. A. R. (1962). Quicksort. *Computer Journal,* **5** (1).

Hockney, R. W. (1965). A fast direct solution of Poisson's equation using Fourier analysis. *Journal of the Association for Computer Machinery,* **12**.

Hockney, R. W. (1986). Parametrization of computer performance. IBM Centre for Scientific and Engineering Computing, Rome, Italy.

Hockney, R. W., and Jesshope, C. (1981). *Parallel computers.* Adam Hilger, Bristol.

Householder, A. S. (1974). *The theory of matrices in numerical analysis.* Blaisdell, New York.

Hyafil, L. and Kung, H. T. (1977). The complexity of parallel evaluation of linear recurrences. *Journal of the Association for Computer Machinery,* **24** (3).

Iliffe, J. K. (1982). *Advanced computer design.* Prentice Hall.

Isaac, E. J., and Singleton, R. C. (1956). Sorting by address calculation. *Journal of the Association for Computer Machinery,* **3** (3).

Jacobi, C. G. J. (1846). Über ein leichtes Verfahren die in der Theorie der Säcularstorungen vorkommenden Gleichungen numerisch aufzulösen. *Journal der reine und angewandte Mathematik,* **30**.

Jacobs, D. A. H. (1980). Preconditioned conjugate gradient methods for solving systems of algebraic equations. Central Electricity Research Laboratory, Note no. RD L N 193 80.

Jennings, A. (1977). Partial elimination. *Journal of the IMA,* **20**.

Jennings, A. (1982). Development of an ICCG algorithm for large sparse systems. In *Preconditioning methods: analysis and applications* (ed. D. Evans). Gordan and Breach.

Jordan, T. L. (1984). Conjugate gradient preconditioners for vector and parallel processors. In *Elliptic problem solvers II* (eds. G. Birkhoff, and A. Schoenstadt). Academic Press.

Kaiser, H. F. (1972). The JK method: a procedure for finding the eigenvectors and eigenvalues of a real symmetric matrix. *The Computer Journal,* **15** (3).

Kalman, R. E. (1960). A new approach to linear filtering and prediction problems. *Transactions of the ASME* Ser. D J. Basic Engrg., 82D.

Knuth, D. E. (1973). *The art of computer programming. Volume* 3: *Sorting and searching.* Addison Wesley.

Kogbetliantz, E. G. (1954). Diagonalization of general complex matrices as a new method for solution of linear equations. *Proceedings of the congress of mathematicians,* **2**.

Kogbetliantz, E. G. (1955). Solution of linear equations by diagonalization of coefficient matrix. *Quarterly Journal of Applied Mathematics,* **13**.

Kogge, P. M. (1973). Maximal rate pipeline solutions to recurrence problems. In *Proceedings of the first annual symposium on computer architecture,* Gainesville, Fl.

Kourouklis, S., and Paige, C. (1981). A constrained least squares approach to the general Gauss-Markov linear model. *Journal of the American Statistical Association*, **76** (375).

Kowalik, J., and Kumar, S. (1982). An efficient parallel block conjugate gradient method for linear equations. *Proceedings of the international conference on parallel processing*.

Kuck, D. J. (1978). *Structure of computers and computations*. Wiley, New York.

Kuck, D. J. (1982). High-speed machines and their compilers. In *Parallel processing systems* (ed. D. Evans). Cambridge University Press.

Kung, H. T., and Stevenson, D. (1977). A software technique for reducing the routeing time on a parallel computer with fixed interconnection network. In *High speed computer and algorithm organization* (eds. D. Kuck, D. Lawrie, and A. Sameh). Academic Press.

Lai, C., and Liddell, H. (1988). Preconditioned conjugate gradient method on the DAP. In *Mathematics of finite elements and applications VI* (ed. J. Whiteman). Academic Press.

Lambiotte, J. (1975). The solution of linear systems of equations on a vector computer. Ph.D. dissertation, University of Virginia.

Laub, A., Heath, M., Paige, C., and Ward, R. (1987). Computations of system balancing transformations and other applications of simultaneous diagonalization algorithms. *IEEE Transactions on automatic control*, AC-**32** (2).

Le Veque, R., and Trefethen, L. (1988). Fourier analysis of the SOR iteration. *IMA Journal of Numerical Analysis*, **8**.

Livesley, R. K. (1983). *Finite elements: an introduction*. Cambridge University Press.

Livesley, R. K., Modi, J. J., and Smithers, T. (1986). The use of parallel computation for finite element calculations. In *Parallel Computing*, **85** (eds. M. Feilmeier, G. Joubert, and U. Schendel). North Holland.

Lootsma, F. A. (1984). Parallel unconstrained optimization methods. Report 84-30, Dept. of Mathematics and Informatics, Delft University, Delft, The Netherlands.

Lord, R., Kowalik, J., and Kumar, S. (1983). Solving linear algebraic equations on an MIMD computer. *Journal of the ACM*, **30**.

Lovelace (Augusta), A., Countess of (1843). Translation of Menabrea's (1842) 'Sketch of the analytical engine invented by Charles Babbage' with editorial notes by the translator. In *Scientific Memoirs* (ed. R. Taylor), III, article xxix. (A reproduction of this paper may be found in Bowden 1953).

Luk, F. T. (1980). Computing the singular value decomposition on the ILLIAC IV. *ACM Transactions on Mathematical Software*, **6** (4).

Madsen, N. K. and Rodrigue, G. H. (1977). Odd-even reduction for pentadiagonal matrices. In *Parallel computers—Parallel mathematics* (ed. M. Feilmeier). North Holland.

Menabrea, L. F. (1842). Sketch of the analytical engine invented by Charles Babbage, Esq. Bibliothèque Universelle de Genève, 82. (A reproduction of its English translation (with editorial notes), by the Countess of Lovelace, may be found in the appendix in Bowden 1953.)

Mitchell, A., and Wait, R. (1977). *The finite element method in partial differential equations.* Wiley.

Modi, J. J. (1982). Jacobi methods for eigenvalue and related problems in a parallel computing environment. Ph.D. thesis, University of London.

Modi, J. J., and Parkinson, D. (1982). Study of Jacobi methods for eigenvalues and singular value decomposition on DAP. *Computer Physics Communications,* **26**.

Modi, J. J., and Bowgen, G. S. J. (1984). QU factorization and singular value decomposition on the DAP. In *Super-computers and parallel computation* (ed. D. Paddon). Oxford University Press.

Modi, J. J., and Clarke, M. R. B. (1984). An alternative Givens Ordering. *Numerische Mathematik,* **43**.

Modi, J. J., Davies, R. O., and Parkinson, D. (1984). Extension of the parallel Jacobi method to the generalized eigenvalue problem. In *Parallel computing,* **83** (eds. M. Feilmeier, G. Joubert, and U. Schendel). North Holland.

Modi, J. J., and Pryce, J. D. (1985). Efficient implementation of Jacobi's diagonalisation method on the DAP. *Numerische Mathematik,* **46**.

Modi, J. J., and Rollett, J. S. (1985). An algorithm for inverse square-roots. *Parallel Computing,* **2**.

Modi, J. J., and Rollett, J. S. (1986). Some problems of exploiting a pipeline processor. *Parallel Computing,* **3**.

Modi, J. J., and Prager, R. W. (1987). Implementation of bubble sort and the odd-even transposition sort on a rack of transputers. *Parallel Computing,* **4**.

Muller, D., and Preparata, F. (1975). Bounds to complexities of networks for sorting and for switching. *Journal of the Association of Computer Machinery,* **22** (2).

Mullis, C., and Roberts, R. (1976). Synthesis of minimum roundoff noise fixed point digital filters. *IEEE Transactions on Circuits and Systems,* CAS-**23**.

Nash, J. C. (1975). A one-sided transformation method for singular value decomposition and algebraic eigenproblem. *Computer Journal,* **18**.

Nassimi, D., and Sahni, S. (1982). Parallel algorithms to set up benes* permutation network. *IEEE Transactions on Computation,* C-**31** (2).

Newcomb, R. W. (1960). On simultaneous diagonalisation of two semi-definite matrices. *Quarterly Journal of Applied Mathematics,* **19**.

O'Leary, D. P. (1984). Ordering schemes for parallel processing of certain mesh problems. *SIAM Journal on Scientific and Statistical Computing*, **5**.

O'Leary, D. P., and Stewart, G. W. (1984). Data-flow algorithms for parallel matrix computations. Tech. report 1366, Computer Science Dept., University of Maryland.

Ortega, J. M., and Voigt, R. G. (1985). Solution of partial differential equations on vector and parallel computers. *SIAM Review*, **27**.

Paige, C. C. (1978). Numerically stable computations for general univariate linear models. *Communications on Statistical and Simulational Computation*, B **7** (5).

Paige, C. C. (1979). Computer solution and perturbation analysis of generalised linear least squares problem. *Mathematics of Computation*, **33**.

Paige, C. C. (1985a). Covariance matrix representation in linear filtering. *Contemporary Mathematics*, **47**.

Paige, C. C. (1985b). The general linear model and the generalised singular value decomposition. *Linear Algebra and its Applications*, **70**.

Paige, C. C. (1986). Computing the generalised singular value decomposition. *SIAM Journal on Scientific and Statistical Computing*, **7** (4).

Paige, C. C., and Saunders (1981). Towards a generalised singular value decomposition. *SIAM Journal on Numerical Analysis*, **18** (3).

Paige, C. C., and Van Dooren, P. (1986). On the quadratic convergence of Kogbetliantz's algorithm for computing the singular value decomposition. *LAA*, **77**.

Parkinson, D. (1981). Sparse matrix vector multiplication on the DAP. Tech. Report, Centre for Parallel Computing, Queen Mary College, London.

Parkinson, D. (1982a). Parallel processing architecture and parallel algorithms. In *New advances in distributed computer systems* (ed. K. Beauchamp), Nato advanced series C, 80. D. Reidel Publishing Company.

Parkinson, D. (1982b). Practical parallel processors and their use. In *Parallel processing systems* (ed. D. Evans). Cambridge University Press.

Parkinson, D. (1987). Organisational aspects of using parallel computers. *Parallel Computing*, **5** (1 and 2).

Parlett, B. N. (1980). *The symmetric eigenvalue problem*. Prentice Hall.

Pease, M. C. (1977). The indirect binary n-cube microprocessor array. *IEEE Transactions on Computing*, **C-26** (5).

Penrose, R. (1955). A generalised inverse for matrices. *Proceedings of the Cambridge Philosophical Society*, **51**.

Preparata, F. (1978). New parallel sorting schemes. *IEEE Transactions on Computing*, **C-27** (7).

Rodrigue, G. H., Madsen, N. K., and Karush, J. I. (1979). Odd-even reduction for banded linear equations. *Journal of the Association for Computer Machinery*, **26** (1).

Rutishauser, H. (1971). The Jacobi method for real symmetric matrices. In *Handbook for automatic computation. Linear Algebra.* (eds. J. H. Wilkinson and C. H. Reinsch), **2**, Springer Verlag.

Saad, Y., and Schultz, M. (1986). Data communication in parallel architecture, research report YALE DCS RR-461, Dept. of Computer Science, Yale University.

Sameh, A. H. (1971). On Jacobi and Jacobi-like algorithms for parallel computers. *Mathematics of Computation*, **25**.

Sameh, A. H. (1985). On some parallel algorithms on a ring of processors. *Computer Physics Communications*, **37**.

Sameh, A. H., and Brent, R. P. (1977). Solving triangular systems on a parallel computer. *SIAM Journal on Numerical Analysis*, **14** (6).

Sameh, A. H., and Kuck, D. J. (1978). On parallel linear systems solvers. *Journal of the ACM*, **25**.

Schendel, U. (1984). *Introduction to numerical methods for parallel computers*. Ellis Horwood Ltd., Chichester.

Siegel, H. J. (1977). The universality of various types of SIMD machine interconnection networks. In *Proceedings of the fourth annual symposium on computer architecture*. ACM SIGARCH IEEE-CS, New York.

Singleton, R. C. (1969). An efficient algorithm for sorting with minimal storage. *Communications of the ACM*, **12**.

Smith, G. D. (1978). *Numerical solution of partial differential equations: Finite difference methods*. Oxford University Press.

Southwell, R. V. (1946). *Relaxation methods in theoretical physics*. Oxford University Press.

Stewart, G. W. (1973). *Introduction to matrix computations*. Academic Press.

Stone, H. S. (1971). Parallel processing with the perfect shuffle. *IEEE Transactions on Computing*, **C-20** (2).

Stone, H. S. (1973). Problems of parallel computation. In *Complexity of sequential and parallel numerical algorithms* (ed. J. Traub), Academic Press.

Strang, G., and Fix, G. (1973). *An analysis of the finite element method*. Prentice Hall.

Traub, J. F. (ed.) (1973). *Complexity of sequential and parallel numerical algorithms*. Academic Press.

Turing, A. M. (1950). Computing machinery and intelligence. *Mind*, **59** (236).

Turner, P. R. (1987). Towards a fast implementation of level-index arithmetic. *Bulletin of the IMA*, **22**.

Van Kempton, H. P. M. (1966). On the quadratic convergence of the special cyclic Jacobi method. *Numerische Mathematik*, **9**.

254   *Bibliography*

Van Loan, C. (1976). Generalizing the singular value decomposition. *SIAM Journal on Numerical Analysis,* **13** (1).

Varga, R. S. (1962). *Matrix iterative analysis.* Prentice Hall.

Wilkinson, J. H. (1962). Note on the quadratic convergence of the cyclic Jacobi process. *Numerische Mathematik,* **4**.

Wilkinson, J. H. (1963). *Rounding errors in algebraic processes.* Prentice Hall.

Wilkinson, J. H. (1965). *The algebraic eigenvalue problem.* Oxford University Press.

Wilson, R. (1972). *Introduction to graph theory.* Academic Press.

Winograd, S. (1975). On the parallel evaluation of certain arithmetic expressions. *Journal of the Association of Computer Machinery,* **22** (4).

Young, D. (1971). *Iterative solution of large linear systems.* Academic Press.

Zienkiewicz, O. C. (1977). *The finite element method.* McGraw-Hill.

Zimmerman, K. (1969). On the convergence of Jacobi process for ordinary and generalised eigenvalue problems. Dissertation No. 4305. Zurich: Eidgenössische Technische Hochschule.

Index